Biological Membranes in Toxicology

Biological Membranes
in Toxicology

Ernest C. Foulkes, Ph.D.
University of Cincinnati
Cincinnati, Ohio

USA	Publishing Office:	Taylor & Francis
		325 Chestnut Street, Suite 800
		Philadelphia, PA 19106
		Tel: (215) 625-8900
		Fax: (215) 625-2940
	Distribution Center:	Taylor & Francis
		47 Runway Road, Suite G
		Levittown, PA 19057-4700
		Tel: (215) 269-0400
		Fax: (215) 269-0363
UK		Taylor & Francis Ltd.
		1 Gunpowder Square
		London EC4A 3DE
		Tel: 0171 583 0490
		Fax: 0171 583 0581

BIOLOGICAL MEMBRANES IN TOXICOLOGY

1 2 3 4 5 6 7 8 9 0

Printed by Edwards Brothers, Ann Arbor, MI, 1998.

A CIP catalog record for this book is available from the British Library.
⊗ The paper in this publication meets the requirements of the ANSI Standard Z39.48-1984 (Permanence of Paper)

Library of Congress Cataloging-in-Publication Data

Foulkes, Ernest C., 1924–
    Biological membranes in toxicology / Ernest C. Foulkes.
        p.   cm.
    Includes bibliographical references and index.
    ISBN 1-56032-762-6 (case : alk. paper)
    1. Molecular toxicology.   2. Cell membranes.   I. Title.
RA1220.3.F68   1998
517.6′4—dc21   98-17546
                                                                                    CIP

ISBN 1-56032-762-6

Contents

Preface

A generation or longer ago it was fashionable to look at the action of hormones, drugs, and poisons through what could be described as "membrane-colored glasses." Somehow, all these agents were suspected, at one time or another, to exert their activity by primarily affecting membrane function. Today we know that this represents an attractive but simplistic view of the nature of cellular responses to such agents. Thus, cytoplasmic receptors, such as that for aryl hydrocarbons, genomic poisons, second messengers, and other intracellular factors have been identified, and all indicate that active sites on the membrane in many instances are not the primary target of physiological messengers (hormones) or of drugs and poisons. Even when functional lesions are observed at biological membranes, the interference with normal function may represent only the end-result of a cascade of reactions, not a primary effect at that location.

For instance, a clearer understanding of solute movement across the membrane clarified the dependence of sodium extrusion on mitochondrial metabolism; the effect of a metabolic poison on Na transport across the membrane would, therefore, be classified as indirect rather than direct. On the other hand, inhibition of Na,K-ATPase by cardiac glycosides represents a direct action on membrane transport of sodium. Similarly, the stimulation of cellular glucose uptake by insulin, at one time ascribed to a direct action on glucose transporters, is now attributed to increased membrane turnover (traffic) leading to the incorporation of additional transporters into the membrane; no direct effect of the hormone on membrane function need be invoked.

The distinction between primary and secondary effects on membrane function may at times be difficult. There is no question, however, that membranes play an important role in many toxicological phenomena. Not only may membrane function be affected by a xenobiotic, but the absorption of this xenobiotic, its distribution in the body and its excretion, and therefore its concentration and half-life at its target site, in many instances are determined by the properties of cell membranes. This conclusion justifies the attempt to provide a general overview of the role played by membranes in toxicology. This effort does not cover the vast field of membrane science, as this has been discussed in great detail in numerous publications. While each chapter includes a historical or background section to permit the reader to place the discussion into a wider scientific context, the emphasis throughout the book focuses on the role of membranes in toxicological phenomena rather than on the underlying membrane science.

This book should be relevant to toxicologists, biochemists, pharmacologists, and others concerned with reactions at the cell and other biological membranes. Its usefulness may be increased by a detailed evaluation of the advantages and limitations of different techniques commonly applied to study membrane function and the mechanism of its alteration by various agents. The treatment of the many facets of membrane toxicology is broad but not necessarily comprehensive. Sufficient key references, however, are cited to document the main conclusions. I apologize to authors whose work I may have overlooked, or whose contributions were among a number not all of which could be quoted.

Ernest C. Foulkes

1

General Introduction

The cellular structure of biologic tissues was not fully recognized until the early nineteenth century, well over 100 years following the introduction of the microscope. The concept of the cell membrane as the structure that, in living cells, separates the cytoplasm from its extracellular surroundings evolved over the same period (see reviews by Kleinzeller [1994, 1996]); these membranes, however, were regarded as no more than passive and semipermeable envelopes shielding cell contents against toxic compounds and pathogenic organisms, and perhaps also reducing the loss of essential metabolites or cytoplasmic proteins.

The identification of the cell membrane as such a protective envelope explained most of the earlier observations. It accounted, for instance, for the fact that damage to the cell membrane can lead to loss of cell constituents. The classic example here is the loss of hemoglobin from red blood cells following their exposure to detergents or hypoosmotic shock, or to a cycle of freezing and thawing. Within certain limits of osmolarity, erythrocytes behave as perfect osmometers; unlike most other cells (see Chapter 8), erythrocytes possess little compensatory ability to counteract constriction or swelling in anisotonic media. Excessive swelling in hypotonic solution leads to hemolysis, leaving behind essentially empty ghost cells. It is interesting to note that under suitable incubation conditions such damaged cells can repair themselves; that is, ghost cells can be resealed.

More modern work has shown that most cells normally can compensate to a significant extent for swelling or contracting in anisotonic media. Such volume homeostasis suggests that cell membranes likely are more than inert and semipermeable envelopes for the cytoplasm. Indeed, they are complex structures, involved in the active or facilitated transport of many solutes; in the determination of cytoplasmic composition and volume; in cellular adaptation, cell recognition, and communication, as the site for receptor interaction with many agonists and antagonists; and in other processes. This structural and functional complexity makes membranes susceptible to attack by many specific or nonspecific toxic agents. Some of the resulting functional lesions reflect primary interaction between membrane constituents and toxicants; others arise secondarily from reactions elsewhere in the cell, as in the case of inhibitors of mitochondrial energy metabolism, which indirectly abolish active solute transport at the membrane. It is the aim of this monograph to explore the nature of such functional membrane lesions and to evaluate, in general, the role played by biologic membranes in the field of toxicology.

The discussion focuses on the mechanism of action of toxicants on membranes, rather than on basic membrane science. Numerous reviews and monographs have been written on normal membrane composition and function, and they provide a detailed background for the present discussion. It seems sufficient for the present purposes, therefore, to provide in each chapter no more than a relatively short summary of the appropriate background, with a few selected references. For instance, the well-known role of cell membranes in immunology is not considered in any depth (section 5-8); the discussion, instead, is restricted to examples of immunotoxic effects exerted on or at the cell membrane.

Although the major emphasis in this work has been placed on the plasma membranes surrounding animal cells, toxic actions on membranes limiting subcellular organelles or compartments also are considered. Effects of toxicants on mitochondrial permeability and volume control, for instance, or on the retention of autolytic enzymes in lysosomes, emphasize the role potentially played by membranes of these organelles in cellular responses to toxic agents.

Protective and sequestering roles of membranes are discussed in Chapter 2. In many instances, as in the case of the blood–brain barrier (see also section 6-7), the relative impermeability of cell membranes protects the underlying tissues against toxicants circulating in blood. This impermeability may reflect the physicochemical properties of the membrane, or it may constitute the net result of active extrusion of xenobiotics from cells. Other toxicants, in contrast, actually may be assisted by membrane mechanisms in reaching their intracellular target sites. The toxicologic significance of cell membranes is further illustrated by their control over absorption and excretion of xenobiotics. Chapter 6 considers the function of membranes in the transport of xenobiotics.

Much of the experimental work on membrane toxicology has been performed with isolated sytems in-vitro. Like all model systems, in-vitro preparations possess advantages and limitations. Such intrinsic limitations have not always been considered adequately, so proposed interpretations of some in-vitro findings are at times open to question. Such questions have been raised, for instance, on the physiologic significance of solute transport, and the toxicologic significance of its inhibition, in slices of renal cortex or in everted sacs of intestine. Chapter 3 discusses the suitability of these and other in vitro preparations for the study of membrane toxicology. The problems arise from such factors as the morphology of the tissue, the polarity of cells, the possible contribution of artifacts, and the problems ultimately posed by the necessity to extrapolate the findings back to the intact organism. As with any reductive approach leading to use of more and more highly simplified models, the implications of in-vitro findings for the whole organism may not always be self-evident. The demonstration, for instance, that an isolated enzyme reacts with a general protein reagent like Hg^{++} gives no indication of whether this enzyme serves as a primary target of Hg^{++} in vivo.

Chapter 4 provides an overview of the basic properties and functions of normal membranes and of their specific constituents such as pores, channels, receptors, recognition molecules, and ectoenzymes. An additional important characteristic of membranes is their fluid and dynamic rather than solid and static nature. Changes in fluidity can alter membrane permeability to toxic and other solutes. Toxic agents also can affect the continuous turnover of membranes and especially the splitting out of fragments from membranes or their insertion. These activities, generally referred to as examples of membrane traffic, are intimately involved in many functions, such as cell movement, exocytosis, and endocytosis, in the cellular response to some hormones and in other functions, as considered in sections 2-6 and 4-3. Cellular responses to changes in the environment also frequently involve modifications in membrane function; such adaptative changes and their sensitivity to toxicants are described in section 7-4.

Recognition of the involvement of membranes in a variety of specific functions focused an early interest on the question of to what extent regulators and toxicants react primarily with the cell membrane. A popular hypothesis some 50 years ago predicted that the primary site of action of many, if not most, hormones and drugs would be found at the cell membrane; similarly, membranes also were suspected of serving as primary targets of many toxic agents. Insulin is a good example of a hormone believed at one time directly to stimulate glucose transport across the cell membrane; recruitment or new synthesis of additional transporters was not suspected under the influence of the hormone. The general hypothesis primarily involving membranes in cellular responses to external stimuli led to extensive work but ultimately proved simplistic, especially with the recognition of second messengers, intracellular receptors, and the genomic action of many agents.

Nevertheless, evidence indicates that membrane reactions do account for the effects of a significant number of nongenomic toxicants, including especially channel blockers, enzyme inhibitors, and receptor ligands (see Chapter 5). Even if toxic agents can be shown to produce functional lesion at the membrane, however, it often proves difficult to distinguish between actions directly exerted at that site and others only indirectly affecting membrane function. Further, the action may nonspecifically alter the function of membranes by affecting their general integrity, for instance by peroxidation of membrane lipids. Alternatively, the toxic action may inhibit specifically, for instance, an enzyme associated with the membrane. The wide range of possible toxic actions on membrane function, whether exerted directly or indirectly, and whether specific or nonspecific, is discussed in Chapter 5.

Extensive research has shown that membranes not only may serve as targets for toxic action but also can facilitate movement of toxic solutes into and out of cells, or even actively accumulate them in cells. In general, the control exerted by membranes on translocation of nonlipophilic xenobiotics is a critical determinant of their toxicokinetic properties. Toxicokinetics include such characteristics as absorption of a toxic agent into the body, its distribution and biologic half-life in the body, its accumulation at its site of action, its cellular uptake and extrusion, and finally its excretion from the body. Obviously the sensitivity of an organ to a toxicant is influenced by the ease with which this agent can penetrate the tissue and by its residence time in a toxic form in the tissue; the residence time is a function of both metabolism and cellular export of the agent. The transport of toxic substances across cell membranes constitutes the topic of Chapters 6 and 7.

Chapter 7 deals with the export of waste products or toxic agents from the cells, as well as with the cellular extrusion of solutes during such processes as intestinal absorption or renal tubular secretion. A number of toxic xenobiotics have been identified that can depress transcellular movement of solutes by inhibiting their extrusion from cells. The export of a toxic agent, of course, tends to reduce its intracellular concentration and thus helps protect cells against its action. If drugs are employed for therapeutic purposes, however, their export results in decreasing cellular sensitivity to drugs, as is seen clearly in the case of antibiotic- or antineoplastic-resistant cells. Inhibitors of toxicant extrusion, to the contrary, can reverse the constitutive or induced drug resistance of cells. The action of such chemosensitizers is of considerable therapeutic significance.

The solute extrusion process represents a constitutive function in many cells, but it may be induced in others in response to exposure. To that extent, we are dealing with an example of functional adaptation at the membrane to the presence of noxious substances. Many such adaptive responses have been observed (see Chapter 7). Adaptation to toxic exposure, of course, also may involve intracellular mechanisms, as when synthesis of the intracellular metal-sequestering protein metallothionein is stimulated by heavy metals.

Finally, Chapter 8 focuses on volume homeostasis and effects of toxicants. Both cellular and subcellular membranes are, indeed, intimately involved both in the maintenance of normal volumes of cells or subcellular organelles and in the compensatory changes permitting return of volumes toward normal steady-state values following acute exposure to anisotonic media. In animal cells, volume homeostasis depends to a significant extent on the active or passive solute permeabilities of cell membranes, rather than on their mechanical ability to withstand hydrostatic pressure. Volume control is therefore found, as expected, to be as sensitive to the action of many inhibitors as is membrane function in general. This correlation is seen clearly in the common observation that cytotoxicity is associated with swollen cells and mitochondria. Cellular edema, indeed, can serve as a useful biomonitor of cytotoxicity.

In summary, membranes participate extensively in the reactions of toxic agents with cells, and their active or passive permeability properties largely determine the toxicokinetics of these agents. Numerous functional membrane lesions have been described in poisoned cells and tissues, whether induced directly at the membrane or indirectly following a primary effect elsewhere in the cell. What emerges from this discussion is the large number of well-documented toxic actions involving biologic membranes. Recognition of such actions justifies the definition of the broad field of membrane toxicology, which provides the major focus for the present volume. To repeat, it is not the aim of the work exhaustively to cover this field. Instead, an attempt is made to document the important role of biologic membranes in toxicologic processes, by providing some of the historical background and by citing and evaluating a sufficient number of critical references. The survey of the literature for this purpose was concluded in the summer of 1997.

REFERENCES

Kleinzeller, A. 1994. The concept of a cell membrane: Early developments. *News Physiol. Sci.* 9: 147–148.
Kleinzeller, A. 1996. William Hewson's studies of red blood corpuscles and the evolving concept of a cell membrane. *Am. J. Physiol.* 271:C1–C8.

2

Toxicological Significance of Biological Membranes

2-1 INTRODUCTION

Cell membranes not only serve in a protective role, as originally suggested, but are also critically involved in many other basic functions. These include uptake of metabolic substrates by cells and the extrusion of waste products, the general maintenance of cell homeostasis, identification of and communication between cells, and the responses to drugs and poisons and to extracellular messengers such as hormones. Underlying these functions is a complex composition and structure that renders membranes directly or indirectly susceptible to many toxic actions. Further, as is discussed here, the toxicokinetic characteristics of nonlipophilic xenobiotics are largely determined by cell membranes. As a result, damage to membrane integrity not only threatens cell composition and volume, and thereby normal function, but also may influence the concentration of toxicants throughout the body.

Functional lesions at cell membranes commonly are associated with cytotoxic action. For example, chlorinated insecticides mentioned in Chapter 5 inhibit transmission of the nerve impulse by interfering with normal sodium and potassium transport across axonal membranes. The close correlation between cytotoxicity and cell swelling (a membrane-related process [see Chapter 8]) further emphasizes the association of functional lesions at the cell membrane with cytotoxicity. Essential roles of membranes also are reflected in the so-called transport diseases such as cystic fibrosis or insulin-dependent diabetes; both of these involve changes in the ability of cell membranes to permit or to mediate transfer of specific solutes into and out of cells. The question of whether cytotoxicity in these and other instances results from primary functional lesions at the membrane or whether membranes are affected secondarily following reactions elsewhere in the body or in the cell is not always easy to answer and is discussed in some detail in Chapter 5.

The toxicologic significance of membranes transcends their role as direct or indirect targets of toxicant action. Indeed, unless a toxicant is sufficiently fat-soluble to be able passively to enter or leave cells, its intestinal absorption, its distribution in the body and ability to reach and maintain a critical concentration at intracellular target sites in specific organs, and finally its renal or hepatic or enteric excretion all are influenced by the properties of cell membranes. To repeat, properties of cell membranes in this manner can determine critically the toxicokinetic characteristics of many xenobiotics.

All these functions contribute to the general toxicologic significance of cell membranes; this constitutes the focus of the present chapter. Other chapters explore in greater detail the mechanisms of toxic interference with membrane function and the transport of toxicants by membranes.

2-2 MEMBRANES AND HOMEOSTASIS

Long before a "steady" state was recognized formally as a nonequilibrium although constant state requiring continuous energy expenditure for its maintenance, common experience indicated that the living organism is not at equilibrium with its surroundings. The constant body core temperature in mammals is perhaps the most obvious example of steady state; it normally sinks toward ambient temperatures only after death. Bernard over 150 years ago clearly realized that, like body temperature, volume and composition of body fluids during life also must be maintained within a small range of acceptable values. Recognition of the importance of steady state was emphasized by his famous generalization that the "constancy of the internal environment is the precondition of independent life." Bernard here was actually referring primarily to the fluid bathing cells, the extracellular fluid, whose volume and composition remain remarkably constant on the basis of the integrative action of the autonomic nervous system and the endocrine glands.

Steady state in the normal organism, of course, must extend beyond the extracellular fluid. The general term *homeostasis*, coined by Cannon in 1927, covers the tendency in higher animals of all body compartments, whether extracellular or intracellular, normally to maintain appropriate steady states of volume, composition, and intensive properties such as temperature and pressure. Numerous compensatory processes based on negative feedback loops indeed have been identified at the levels of both extracellular and intracellular compartments, and many of these maintain homeostasis on the basis of membrane activities.

Membrane permeability and transport processes, for instance, define the steady-state electrolyte activities in the cytoplasm and consequently the magnitude of transmembrane electric potentials. Function of excitable cells would be compromised if the resting activity gradients of sodium and potassium across the cell membrane did not remain stable within narrow limits. One may recall in this connection that local anesthetics block nerve conduction by altering the electrolyte permeability and consequently the transmembrane electric potential of the nerve axon. The magnitude of transmembrane activity gradients that must be maintained to permit normal function is generally high, both in excitable and in nonexcitable cells; a well-known example is the parietal cell in the gastric mucosa, which establishes and maintains H^+ gradients approaching $10^6:1$.

Cells can maintain a constant volume even in the face of significant osmotic insults, and to a large extent do so by virtue of compensatory changes in membrane function (see Chapter 8). The importance of such compensatory activity for volume homeostasis is especially critical for cells of renal collecting ducts and the urinary bladder. These may be exposed to highly dilute or concentrated solutions, ranging in rats from below 200 mosM to perhaps as high as 3000 mosM. Compensatory volume contraction or expansion counteract the influence of these extreme variations on extracellular osmotic pressures. The major mechanisms contributing to this compensation are the regulation of solute fluxes across the cell membrane and the metabolism of organic osmolytes (see Chapter 8); the compensation is sensitive to inhibition by toxic agents.

Not all solutes in cytoplasm, of course, contribute significantly to its osmolarity. Nevertheless, normal cellular function requires that their concentrations also be maintained within narrow limits. A prime example is ionized Ca in the cytoplasm. Any interference with normal uptake or extrusion of this ion across the cell membrane, or with its sequestration or release from subcellular stores, generally leads to cytotoxic effects. This correlation between alterations in cytoplasmic levels of ionized Ca and cytotoxicity has been observed frequently.

Membranes determine solute fluxes into and out of cells both on the basis of both their passive permeability properties and the presence of a variety of facilitated-diffusion or active-transport mechanisms. The negative feedback loops that control these processes, and thus regulate many homeostatic compensatory reactions, depend on appropriate permeability changes of membranes and also, in some cases, utilize membranes as the site of sensing mechanisms. For instance, swelling of cells in hypotonic media, with concomitant stretching of the membrane, directly alters the activity of stretch-sensitive ion channels, an important compensatory reaction contributing to maintenance of normal cell volumes (see Chapter 8). Presence of voltage-gated ion channels permits membranes of excitable cells to maintain high transmembrane potentials until activation by a depolarizing action potential. Opening and closing of the sodium and potassium channels thereupon permit rapid reestablishment of normal steady state electrolyte gradients.

2-3 MEMBRANES AS TARGET SITES OF TOXIC ACTION

The active involvement of membranes in many specialized processes implies, as discussed previously, a likely sensitivity of membrane function to the action of various toxic agents. For instance, a general protein reagent like mercuric mercury directly and nonspecifically affects many membrane reactions. As a corollary of such a finding, it follows that the inhibition of an isolated enzyme in vitro by this metal cannot automatically identify the primary toxicologic target of this agent in the intact organism. This important conclusion is referred to repeatedly in other chapters.

Very broad toxic effects may be elicited directly at the membrane by treatments that induce lipid peroxidation and thus threaten the general integrity of the lipoprotein structure. Similarly, the importance of phospholipids in membrane structure accounts for the ability of phospholipases nonspecifically to affect membrane processes. Note that both these examples refer to nonspecific actions exerted directly at the membrane (see further section 5-4).

In contrast to the general actions of these agents, other xenobiotics may directly and specifically inhibit one of the many specialized functions carried out at the membrane. A classical example is the cardiac glycoside inhibition of Na,K-ATPase, an enzyme involved in active sodium extrusion across the cell membrane. Because such inhibition of ion transport alters cell composition, and because function of a cell closely depends on its composition, depression of this ATPase activity is likely to be sensitively reflected in altered cell function. Similarly, presence of specific receptor proteins on the membrane makes them likely and direct targets for receptor antagonists, with immediate implications for cellular responses to extracellular stimuli.

Finally, membranes also may serve as indirect targets of toxic agents. In this instance, some reaction elsewhere in the body or in the cell affects membrane processes. An obvious example is the inhibition of oxidative phosphorylation in mitochondria, with consequent depression of active membrane transport. Representative examples of such direct and indirect actions on the membrane are considered in Chapter 5.

2-4 PROTECTIVE AND SEQUESTERING ROLES OF MEMBRANES

A dying or dead cell loses its permeability barrier at the membrane. Thus, loss of viability is accompanied by, for example, loss of dye exclusion and leakage of cytoplasmic contents. Thus, the appearance of the cytoplasmic enzyme lactate dehydrogenase in extracellular fluid is a sensitive biomarker of cell damage. In contrast, normal cell membranes minimize loss by diffusion of many essential cell constituents. The same impermeability also helps protect critical intracellular sites against toxic actions of extracellular agents. Such protection may further involve the activity of membrane carrier systems mediating extrusion of toxic compounds from cells (see Chapter 7); this, in effect, reduces the effective permeability of the membrane. Finally, as described in section 4-1, effects of extracellular toxicants may be minimized by the presence of membrane-associated mucous layers; such mucus may itself become a target of toxic action.

In terms of the whole organism, cell membranes may reduce access of toxicants to their targets by acting as barrier to their absorption from the intestine or their reabsorption from the glomerular filtrate in the kidney. Renal or enteric secretion, another function of epithelial cell membranes, also reduces systemic concentrations of toxicants. This is considered in greater detail in section 6-2 under the heading of the role of membrane transport mechanisms in toxicokinetics.

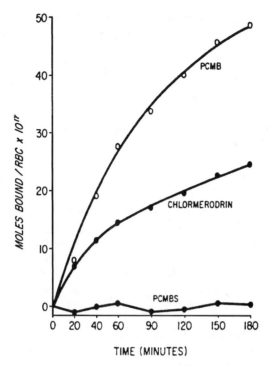

FIG. 2-1. Binding of mercurials by erythrocytes. (From Vansteveninck, Weed, and Rothstein, 1965; reproduced from *J. Gen. Physiol.* by copyright permission of the Rockefeller University Press.)

The protective role of membranes at the level of potential target cells is well illustrated by the results of Vansteveninck, Weed, and Rothstein (1965) on the localization of sulfhydryl groups in erythrocyte membranes. As shown in Fig. 2-1, the membrane in the intact cell imposes a diffusion barrier: Chlomerodrin, a mercurial diuretic, and p-chlormercuribenzoate slowly penetrate the cells; little p-chlormercuribenzenesulfonate is taken up. Nevertheless, both of the latter two strongly and to a similar extent inhibit the uptake of glucose, presumably by reacting with critical sulfhydryl groups on the external membrane. P-chlormercuribenzoate however, reacts much faster with constituents of hemolyzed than of intact cells, an indication that access of this mercurial to the cell interior normally is reduced by an intact membrane. This recalls the dye exclusion test frequently used to monitor cell viability.

Cell membranes acting as diffusion barrier against toxicants protect intracellular targets only against nonlipophylic agents, normally unable passively to penetrate the cell; little protection is offered against lipid-soluble toxicants. Unable in the absence of special channels or transport mechanisms to cross the membrane, the lipid-insoluble agents may interact at least transiently with membrane constituents during their transmembrane movement. Such reactions have been demonstrated, for instance, for heavy metals (see section 6-3). Thus, in rat jejunum luminally perfused in situ with saline containing low metal concentrations, Cd^{++} and some other polyvalent cations bind to fixed anionic sites on the brush border as a first step in their cellular uptake (Foulkes, 1985). High concentrations of the cations inhibit each other's binding, presumably by neutralizing the fixed charges. In contrast, as further discussed in section 6-3, Hg^{++} is present in the perfused lumen primarily as a polychloride anion, and its uptake is stimulated, not inhibited by, for example, La^{3+}. Presumably the neutralization of fixed anionic charges by the polyvalent cation facilitates the approach to the membrane of diffusible anions such as the trichloride complex of Hg^{2+}.

The likelihood that the membrane barrier can protect a cellular target site is diminished, of course, if this site is associated with the membrane itself. Little or no protection can be expected for a target on the outer surface of the membrane. Some shielding may occur if the active site subject to inhibition is deeply buried in the membrane or is located on its inner surface. A possible example of the inaccessibility of a membrane site to an extracellular toxicant is cited in section 3-7 in relation to the Cd sensitivity of amino acid transport in cell-membrane vesicles.

In the cases of a number of toxicants the protective or sequestering function of cell membranes is replaced by the opposite role, the perhaps fortuitous transport of certain toxicants into cells (see Chapter 6). Use of the word "fortuitous" here refers to the likelihood that the toxic agent may be utilizing a transport mechanism evolved for other purposes. The nephrotoxicant cephalosporin, for instance, makes use of the organic anion transporter in the renal tubule; paraquat crosses the basolateral cell membrane with the aid of the organic cation transport system (see section 6-6). During such transfer across a membrane the toxicant is likely to react with membrane constituents, as discussed above for heavy metals, and this binding, although transient, temporarily at least may alter membrane function. Presence of bound metals, for instance, may affect function by altering electrical-charge distribution on the membrane and consequently the transmembrane potential.

The impermeability that permits membranes to minimize cellular uptake of many xenobiotics also helps avoid, as previously stated, the significant loss by diffusion of important metabolites and cell constituents. Finally, membranes also serve to sequester toxic entities or enzymes within cells, or within subcellular organelles like peroxisomes or lysosomes. An example of such sequestration is the mucosal trapping of Cd as Cd-metallothionein (CdMT) during jejunal Cd absorption, discussed below. Another instance is the hepatic accumulation of Cd-metallothionein following injection or oral administration of inorganic Cd. Complexed as Cd-metallothionein the metal generally is believed to be relatively nontoxic and is not readily diffusible across hepatic or enteric cell membranes. Damage from any source to the hepatocytes, however, permits increased leakage of CdMT into plasma, followed by its preferential uptake into renal tubular epithelium (see section 6-6). The mechanism by which such uptake leads to nephrotoxicity remains under discussion.

The ability of endogenous MT to trap Cd up to some saturation level in the jejunal mucosa is illustrated in Fig. 2-2, in which induction of metallothionein by pretreatment with Zn is seen to increase the ability of mucosal cells to retain Cd taken up

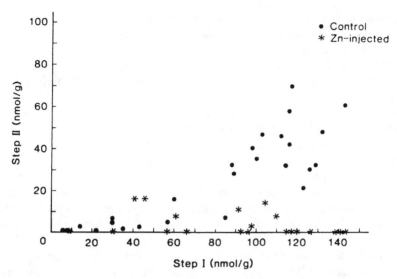

FIG. 2-2. Endogenous metallothionein as determinant of jejunal Cd absorption in the rat. (From Foulkes and McMullen, 1986; with kind permission from Elsevier Science Ireland, LTD, Bay 15K, Shannon Industrial Estate, Co. Clare, Ireland.)

from the lumen (Foulkes and McMullen, 1986). In these experiments Cd in saline was circulated through jejunal segments of the rat in situ. Note that in control rats significant amounts of the Cd extracted from the perfusate (step I of absorption) escaped trapping and were extruded further into the body (step II) only if mucosal accumulation exceeded about 80 nmol/g. In contrast, after Zn pretreatment, step II remained insignificant even if Cd transported by step I reached 140 nmol/g. The pretreatment had been shown previously to increase greatly the mucosal concentration of metallothionein (Bonewitz et al., 1982). These results illustrate an instance of cellular sequestration of a toxicant. Much of the sequestered metal ultimately is lost from the body on normal desquamation of the cells, without ever being fully absorbed.

As another illustration of the sequestering role of membranes, reference may be made to the well-known fact that damage to the lysosomal membrane can lead to escape of highly active proteolytic and other enzymes into the cytoplasm, frequently with resultant cell death. Such a role of the lysosomes is considered in section 6-6 specifically as plausible explanation for the nephrotoxicity of CdMT. Section 5-7 provides additional information on the topic of subcellular membranes in toxicology.

In spite of the inability of nonlipophilic xenobiotics passively to cross cell membranes, their transfer across cellular barriers in general seems to follow pathways across cells rather than through tight junctions between cells; this usually can be deduced a priori from such findings as structural specificity and the occurrence of competitive inhibitors. Such barriers are the intestinal mucosa, the renal tubular epithelium, the blood–brain barrier, the placenta, blood–follicle barriers in the ovaries, the bladder epithelium, and others. It follows that the movement of these solutes during such processes as intestinal absorption, renal tubular transport, and cerebral or fetal solute uptake largely are determined by the properties of cell membranes. In each case, the selective impermeability of cell membranes constitutes a critical factor helping to control access of toxicants to the underlying tissue; at the same time, of course, their impermeability protects function of the barrier cells themselves.

In addition to membrane function, other factors also may contribute to the control of solute transfer across cell barriers. In particular, the process may be slowed, if not largely abolished, by intracellular sequestration or trapping of the solute. This is illustrated by the previous example of endogenous metallothionein-reducing absorption of Cd in the rat jejunum.

The variable ability of the blood–brain barrier to exclude Hg and its compounds from the brain provides a useful illustration for many of these conclusions. Mercuric mercury circulating in plasma either as a nondiffusible complex with plasma protein or as a hydrophilic complex with diffusible ligands such as cysteine or glutathione does not cross to an appreciable extent the blood–brain barrier. In contrast, elemental mercury (mercury vapor) or monomethylmercury more readily is taken up into the brain. This results, at least in part, from their significant lipid solubility. An active transport mechanism further contributes to uptake of monomethylmercury complexed with L-cysteine (see section 7-7). In contrast, and as expected, the lipid nature of the endothelial membranes (see Chapter 4) leaves them with little ability to control transfer of lipophilic substances into the brain. This is illustrated dramatically by the very high neurotoxicity of dimethylmercury, a highly lipid-soluble compound of mercury.

Advantage can be taken of the lipid composition of membranes in attempts to improve drug delivery to tissues; the same procedures also can increase the uptake of toxic xenobiotics. For instance, permeation of a test solute can be stimulated by raising its lipid solubilitiy, for instance by conjugation with a fatty acid or by other suitable substitutions. Finally, the use of liposomes as drug carriers depends on their ability to fuse with the lipid matrix of plasma membranes. The importance of lipophilicity in drug action and toxicology is reviewed in some detail in a recent monograph (Pliska, Testa, and Van der Waterbeemd, 1996). Inversely, detoxification of xenobiotics by conjugation with glycine, glucuronic acid, or other water-soluble molecules reduces the likelihood of their renal reabsorption by taking advantage of the generally low passive permeability of cell membranes to nonlipophilic compounds.

Movement of many solutes across cell barriers, as already mentioned, follows primarily transcellular pathways and does not involve extensive passage through tight junctions and intercellular spaces; both apical and basolateral cell membranes therefore are likely to help determine the rate of transfer of these solutes into the underlying tissue, or into renal tubular fluid, bile, and so forth. Section 7-1 discusses the problem of deciding whether similar or different mechanisms are responsible for the transfer of a solute across the two opposing cell membranes. Evidence obtained in vivo for apparent differences in drug susceptibilities of solute transport at the two sides of a cell barrier must be interpreted with caution. For example, the concentration of a toxic compound at the apical cell membrane in the renal tubule need not be the same as that on the basolateral side. This fact renders it difficult to compare the sensitivities of transporters on the two sides to a given inhibitor in vivo.

In the case of renal tubular reabsorption of amino acids, the basolateral and apical substrate specificities are similar. Inhibitor sensitivities of membrane vesicles isolated from the two sides differ, however, as detailed in section 7-1. On the other hand, transport of the amino acid analogue cycloleucine is inhibited by mercury on both sides of renal tubule cells. As far as transfer of metals themselves is concerned, their transepithelial movement, to the extent that they are not sequestered in the cells, can be attributed to quite different mechanisms at apical and basolateral membranes. Although little direct evidence is available for the nature of metal extrusion, it appears to consist of passive movement of diffusible metal complexes, whereas apical uptake requires metal binding sites on the membrane (see Chapter 6).

An important factor that may increase the permeability of cell membranes, and thereby decrease their protective role against toxicants, is membrane fluidity. For example, Aungst (1996) described the enhancement of solute permeation across cell membranes accompanying an increased fluidity of cell barrier lipids. Raising the fluidity of brush border membranes with n-butanol caused a several-fold increase in the initial rate of Cd penetration into brush border membrane vesicles from rat jejunum (Bevan and Foulkes, 1989); this is illustrated in Fig. 2-3. The stimulating effect of n-butanol on metal uptake, however, is not specific for metals: The membrane becomes more permeable also to other solutes. For instance, the alcohol also made

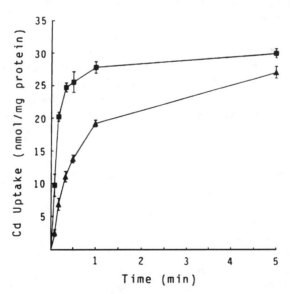

FIG. 2-3. Effect of n-butanol on Cd uptake by membrane vesicles of jejunal brush border of the rat. *Triangles* indicate control vesicles; *squares* mean in presence of 1.5% (v/v) n-butanol. (From Bevan and Foulkes, 1989; with kind permission from Elsevier Science Ireland Ltd., Bay 15K, Shannon Industrial Estates, Co. Clare, Ireland.)

intravesicular cadmium more accessible to the chelotor ethylene diamine tetraacetic acid (EDTA) in the medium (results not illustrated here); normally EDTA is not taken up into the vesicles.

The proposed role of membrane fluidity in the movement of heavy metals across membranes is discussed in greater detail in section 6-3. Both intrinsic and extrinsic factors such as age and diet may influence the fluidity and other membrane properties and thereby affect the ability of the membrane to protect the cell or the body against toxic agents.

In summary, the relative impermeability of cell membranes protects intracellular targets against many toxic xenobiotics and reduces the likelihood of important cellular solutes being lost from the cell. It also limits the passive movement of lipophobic solutes across cell barriers, and thereby plays an important role in determining toxicokinetics of such agents (see section 6-2). The contribution of cell membranes to toxicokinetics becomes especially important if they catalyze transfer of xenobiotics by active or facilitated transport. Such mechanisms may not only stimulate, for example, intestinal absorption of toxic agents but in selected cases also may ease their access to the cell interior (see section 1-5).

An important additional contribution to the protective function of cell membranes derives from the existence or adaptive synthesis of membrane mechanisms able to extrude unwanted solutes from the cell. Such processes can account for instances of bacterial resistance to antibiotics (Nikaido, 1996) or the development of multidrug resistance (Gottesman and Pastan, 1993) in tumor or other cells; Chapter 7 may be consulted for further details. Additional membrane-related processes reduce cytoplasmic activities of toxic agents: One well-studied example is the clearance of Ca^{++} from cytoplasm, both by uptake into subcellular organelles and by extrusion across the cell membrane.

Selective membrane permeability, with exclusion of many extracellular solutes from the cells, is characteristic only of living cells. This fact provides the rationale for using dye exclusion from cells as measure of their viability. It also reemphasizes the significance of the correlation between the integrity of cells and their selective permeability.

2-5 MEMBRANES AND THE DISTRIBUTION OF TOXICANTS IN THE BODY

Not only are cell membranes frequently the site of toxic lesions, but they also participate extensively in the transport of toxicants. Membrane-mediated uptake of toxicants by cells is discussed in Chapter 6, and Chapter 7 covers extrusion from cells. The extent to which, in general, membranes control uptake, distribution, and excretion of xenobiotics is discussed in section 6-2. What clearly emerges is the critical role played by membranes in the toxicokinetics of nonlipophilic agents. External exposure of an organ to such agents can be monitored by analyzing the curve of their plasma concentrations against time. In turn, the shape of this curve is the resultant of activities at many membranes: absorption through intestinal and other cell membranes, cellular uptake and possible sequestration or metabolic breakdown in target organs, and finally excretion by the renal or another route.

The important role of membranes in controlling toxicokinetics becomes particularly obvious if membrane transport mechanisms actually are responsible directly for mediating intracellular accumulation of toxic agents to critical levels in target cells. Facilitation of toxicant uptake by cells is discussed in greater detail in Chapter 6. The protective function of membranes in facilitating export of toxicants from cells is considered further in Chapter 7. The topics are raised here only to contrast them with the original and simple concept of a passive protective role of membranes, and thereby to emphasize the complex influence of membranes on toxic phenomena.

2-6 TOXIC ACTION ON MEMBRANE SYNTHESIS AND TRAFFIC

Plasma membranes are dynamic structures, continuously breaking and fusing during processes such as endo- and exocytosis, penetration of viruses into cells, cell movement, and cell division. They also may undergo compensatory and adaptive changes in response to acute or chronic challenges and are influenced by physiologic factors such as age, development, and diet. Physiologic messengers may increase membrane turnover, as does insulin in response to hyperglycemia (see below). All these reactions constitute the phenomenon of membrane traffic. Section 4-3 provides further details about the lability of membrane structure.

Compensatory changes in membrane function, elicited in response to acute extracellular stress, may involve simply, for instance, opening of existing solute channels. Such a response can be illustrated by stretch-sensitive K channels in osmotically swollen cells (see Chapter 8); there is no need to invoke changes in membrane traffic in this action. In contrast, hyperglycemia elicits secretion of insulin, which, in turn, stimulates membrane traffic. This traffic causes relocation of glucose transporters from within the cell to the cell membrane. Hirshman et al. (1990), for instance, observed that the number of glucose transporters in the plasma membrane of rat skeletal muscle, as measured by the binding of cytochalasin B, was doubled under the influence of insulin; simultaneously the number of glucose transporter molecules in the microsomal fraction was decreased.

More long-range adaptive responses to altered environments may require the synthesis of additional carriers to be incorporated into the cell membrane. Accordingly, agents depressing protein synthesis, such as oligomycin, may interfere with adaptation. Compensatory and adaptive membrane responses to changes in the extracellular medium are further considered in Chapters 7 and 8.

Like other active membrane processes, membrane traffic may serve as a target for toxic agents. A proposed instance of such interference with normal membrane traffic is found in the work of Forge and Richardson (1993). These authors suggest that the ototoxicity of neosporin is related to the incorporation of abnormal membrane into the apical surface of cochlear hair cells. The difference between normal and neomycin-affected membranes was noted indirectly in freeze–fracture studies.

Another example of inhibition of membrane traffic is the effect of the ionophore monensin on pinocytosis in rat-embryo fibroblasts (Wilcox, Kitson, and Widnell, 1982). Of course, to the extent that membrane recycling is an energy-requiring process, it also can be nonspecifically depressed by metabolic inhibitors.

A number of endocytosis and transcytosis inhibitors have been described, as reviewed by Hastings et al. (1994). These authors studied the effects of monensin and nocodazole and found them to disrupt cellular membrane traffic in the lung. Spiro et al. (1996) focused on the effects of wortmannin on the transferrin-receptor pathway in vivo and in vitro. Wortmannin is a fungal metabolite and inhibitor of phosphatidylinositol 3–kinase. In as low as nanomolar concentrations, it irreversibly blocks the insulin-stimulated translocation of vesicles containing glucose transporters and their fusion with the cell membrane (Clarke et al., 1994). The compound was shown to act at the signalling step that triggers translocation of the transporters.

Further information on the nature of membrane traffic and its inhibition by toxic agents is contributed by Frost and Lane (1985), who reported that phenylarsine oxide inhibits the insulin stimulation of translocation of glucose transporters to the plasma membrane of 3T3-L1 adipocytes; vicinal sulfhydryl groups appear to be involved in this translocation. Protein inhibitors of membrane traffic also have been described (Boman and Kahn, 1995). It must be pointed out that inhibition of bacterial cell-wall synthesis by penicillin antibiotics does not involve the plasma membrane and is not covered, therefore, in the present review.

Section 4-3 further emphasizes the qualitative and quantitative significance of membrane traffic. The present chapter leads to the important conclusion that adverse membrane effects of toxicants may result not only from interference with normal function of existing membranes but also from depression of membrane synthesis and turnover. In the absence of normal membrane traffic cells may not be able to grow and divide, to take up or to secrete various macromolecules, to respond to normal physiologic commands such as increasing glucose uptake in presence of insulin, or to adapt, as discussed in Chapter 7, to changes in the diet or other environmental variables.

REFERENCES

Aungst, B. 1996. Oral mucosal permeation enhancement. In *Oral mucosal drug delivery*, ed. M.J. Rathbone, 65–83. New York: Marcel Dekker.

Bevan, C., and E.C. Foulkes. 1989. Interaction of cadmium with brush border membrane vesicles from the rat small intestine. *Toxicology* 54:297–309.

Boman, A.L., and R.A. Kahn. 1995. ARF proteins: The membrane traffic police. *Trends Biochem. Sci.* 20:147–150.

Bonewitz, R.F., E.C. Foulkes, E.J. O'Flaherty, and V. Hertzberg. 1982. The effects of dexamethasone on the kinetics of jejunal zinc uptake and metallothionein synthesis in the rat. In *Biological roles of metallothionein*, ed. E.C. Foulkes, 202–214. Amsterdam: Elsevier.

Clarke, J.F., P.W. Young, K. Yonezawa, M. Kasuga, and G.D. Holman. 1994. Inhibition of the translocation of GLUT1 and GLUT4 in 3T3-L$_1$ cells by the phosphatidylinositol 3–kinase inhibitor, wortmannin. *Biochem. J.* 300:631–635.

Forge, A., and G. Richardson. 1993. Freeze fracture analysis of apical membranes in cochlear cultures: Differences between basal and apical-coil outer hair cells and effects of neomycin. *J. Neurocyt.* 22:854–867.

Foulkes, E.C. 1985. Interactions between metals in rat jejunum: Implications on the nature of cadmium uptake. *Toxicology* 37:117–125.

Foulkes, E.C., and D.M. McMullen. 1986. Endogenous metallothionin as determinant of intestinal cadmium absorption: A reevaluation. *Toxicology* 38:285–291.

Frost, S.C., and M.D. Lane. 1985. Evidence for the involvement of vicinal sulfhydryl groups in insulin-activated hexose transport by 3T3-L$_1$ adipocytes. *J. Biol. Chem.* 260:2646–2652.

Gottesman, M.M., and I. Pastan. 1993. Biochemistry of multidrug resistance mediated by the multidrug transporter. *Ann. Rev. Biochem.* 62:385–427.

Hastings, R.H., J.R. Wright, K.H. Albertine, R. Ciriales, and M.A. Matthay. 1994. Effect of endocytosis inhibitors on alveolar clearance of albumin, immunoglobulin G, and SP-A in rabbits. *Am. J. Physiol.* 266:L544–L552.

Hirshman, M.F., L.J. Goodyear, L.J. Wardzala, E.D. Horton, and E.S. Horton. 1990. Identification of an intracellular pool of glucose transporters from basal and insulin-stimulated rat skeletal muscle. *J. Biol. Chem.* 265:987–991.

Nikaido, H. 1996. Multidrug efflux pumps of gram-negative bacteria. *J. Bacteriol.* 178:5853–5859.

Pliska, V., B. Testa, and H. Van der Waterbeemd. 1996. *Lipophilicity in drug action and toxicology.* New York: VCH Publishers.

Spiro, D.J., W. Boll, T. Kirchhausen, and M. Wessling-Resnick. 1996. Wortmannin alters the transferring receptor pathway in vivo and in vitro. *Mol. Biol. Cell* 7:355–367.

Vansteveninck, J., R.I. Weed, and A. Rothstein. 1965. Localization of erythrocyte membrane sulfhydryl groups essential for glucose transport. *J. Gen. Physiol.* 48:617–632.

Wilcox, D.K., R.P. Kitson, and C.C. Widnell. 1982. Inhibition of pinocytosis in rat embryo fibroblasts treated with monensin. *J. Cell Biol.* 92:859–864.

3

Experimental Approaches

3-1 INTRODUCTION

A variety of experimental models have been employed to study the effects of toxicants on membrane function; they range from the intact animal to artificial membranes. The relatively simple control over experimental variables in vitro, the desire to minimize use of live animals in toxicologic testing and research, and factors of cost and convenience, however, have encouraged an increasing reliance on in-vitro preparations. Moreover, the complexity of the whole animal may defeat attempts to study details of biologic mechanisms, frequently leaving little alternative to reliance on in-vitro models. On the other hand, in-vitro preparations demonstrate what toxicants can affect, not necessarily what they do affect in the more complex setting of an intact animal.

Obviously, whether intact animals or isolated systems are preferred, the design of experimental studies must insure, as always, that the questions raised can be answered logically and unambiguously with the experimental preparation and techniques to be employed. This basic fact may be restated in the present context by insisting that results obtained from the analysis of membrane function in an experimental model cannot be appropriately interpreted if the intrinsic limitations of this model are not appreciated, and if physiologic reality is ignored. Much has been written on this general topic, as well as specifically on how it relates to the study of membrane function and toxicology (see, e.g., Foulkes, 1996).

19

Every experimental model possesses intrinsic or artifactual advantages and limitations. If these limitations are not taken into account adequately, interpretation of experimental findings achieved may remain open to serious question. This is illustrated here with special emphasis on two preparations that, in spite of reservations repeatedly expressed about their general applicability, continue to be commonly employed for the study of membrane transport and its inhibition by toxic agents.

The two model preparations are thin slices of the renal cortex and everted sacs of the small intestine. Both avoid use of living animals and minimize the number of required animals and are therefore economical. They also are relatively simple and convenient to use and make it easy to obtain replicate results. They can yield unambiguous answers, however, only to selected questions on membrane function and its inhibition (see sections 2-3 and 2-5). Although much important knowledge has been gained with their use about processes such as tissue metabolism, certain solute transport processes at cell membranes, and the action of some inhibitors, their uncritical application to all kinds of problems has led to inappropriate conclusions. Tissue slices, in particular, have been studied extensively for many years. Their popularity was further enhanced following the development of greatly improved and standardized cutting and handling procedures; the product of these improvements was described as "precision-cut" slices (Ruegg, Gandolfi, Nagle et al., 1987). Although these possess the advantages of greater uniformity and functional stability than older slice preparations, they cannot escape the intrinsic limitations of slices in general.

The widespread application of slices, no matter how prepared, has been accompanied at times by neglect of these limitations. As a result, physiologic reality has been ignored repeatedly, and empirical findings misinterpreted. The problems created by excessive reliance on renal cortical slices, for instance, are illustrated clearly by the frequent but inappropriate identification of solute uptake from the medium with the process of proximal tubular reabsorption. This problem is considered in greater detail in section 2-3.

Other in-vitro preparations commonly employed in the study of membrane function and its inhibition run the gamut of complexity from whole perfused tissues to artificial lipid membranes incorporating specific membrane constituents. The range includes tissue fragments like collagenase digests of renal tubules, slices, ring-like sections of intestine or blood vessels, cell cultures, membrane patches, and purified membrane vesicles. Measurement of solute transfer in isolated and perfused single sections of renal tubules, for instance, and the effects of toxicants on such transfer have contributed greatly to research in membrane toxicology. This chapter presents a general analysis of the advantages and shortcomings of the more common in-vitro preparations for the study of membrane function and its inhibition by toxic agents.

3-2 GENERAL PROBLEM

Reliance on ever more simplified in-vitro models for biologic research in general, and for the study of toxicologic processes in particular, not only must be predicated on full recognition of the morphology and functional characteristics of isolated tissues, cells, or subcellular fragments but also must recognize the difficulty of extrapolating back to the intact organism. Indeed, extrapolation may not be a linear function consisting of simple additions and subtractions (see, e.g., Schultz, 1996). As a result, the uncertainties unavoidably attaching to these extrapolations are increased as toxicologic reactions are studied in more and more purified preparations.

An extreme example of this problem is the question, repeatedly raised in this book, of whether demonstration of the inhibition of an isolated enzyme by a toxic agent can identify the target of this agent in the intact organism; a precondition for such an identification is the finding that the inhibition is specific for this enzyme. It is not likely, for instance, that inhibition of Na,K-ATPase by a general protein reagent like Hg^{++} in vitro points to this enzyme as a critical site of action of the metal in vivo. This problem is further considered below.

A perhaps somewhat less extreme case is the move from intact animals to isolated membrane vesicles (see section 3-7). The possibility of the loss of membrane constituents or control mechanisms during preparation, however, and the altered access of substrates, messengers, and toxicants to the sites of their metabolism or action, may lead to different results at the two levels of complexity.

Techniques of in-vitro toxicology undoubtedly will continue to provide important information on membrane function and its inhibition by toxic agents. It is not, however, the purpose of this chapter to consider the general usefulness of alternatives to the use of intact animals in the study of toxicology. Instead, attention focuses on the suitability of specific simplified models frequently employed in studies of membrane toxicology. Choice of the optimum model for mechanism-oriented studies in this field must be governed, as always, by at least five important considerations, in addition to the desire to minimize pain and reduce costs.

Experimental Convenience and Control

Obviously, exposure of cells to precisely determined and relatively constant concentrations of toxicants can be more readily achieved with cultured cells than in whole tissues or in intact animals. In addition, it is likely that higher toxicant concentrations can be used in vitro than are compatible with life of an intact organism. Further, speciation of a toxicant such as one of the heavy metals can much more readily be controlled in vitro than in vivo.

Possibility of Artifacts

Preparation of a simplified model system may introduce artifacts or lead to loss of important properties. In epithelial cell suspensions, for instance, loss of cell polarity may provide a biased picture of membrane function. Another example of the loss of a specific function associated with the preparation of an in-vitro model was reported by Liu, Liu, and Klaassen (1994): the ability of renal tubule cells to take up

Cd-metallothionein in vivo was not maintained by cultured cells in vitro. Another type of artifact may be created by the fact that feedback loops normally controlling cell homeostasis may be inoperative in an isolated system; an example is the loss of contact inhibition in cell suspensions.

Structural Limitations

These may be illustrated by the extent to which the collapse of tubules in renal cortical slices (see section 3-3) abolishes the usefulness of this particular model system for the study of transport processes at the apical (luminal) cell membrane. Another example of this problem is the long diffusion path of absorbed solute through the submucosal tissue in everted sacs of the small intestine, coupled with possible trapping of some absorbed solutes in that tissue. This has been observed, for instance, with heavy metals (see section 3-5). Further, the assumption that an absorbed solute exclusively follows a transcellular rather than a pericellular pathway under physiologic conditions and in vitro is not being tested routinely. Such intrinsic structural and functional limitations of in-vitro models always must be adequately taken into account.

Differences in the Ability of Toxicants to React at Their Target Sites in Vivo and in Vitro

This not only poses a problem in terms of physical access, as discussed for the proximal brush border in renal cortical slices, but relates also to other factors. One of these is the speciation of the toxicant. For instance, as referred to previously, a heavy metal circulating in plasma may not be present in the same chemical form as it is in an incubation medium in vitro; cells in vivo generally are not exposed to free heavy-metal ions, nor necessarily to the same toxicant concentration on all sides. Another complicating factor is that cells or fragments of certain tissues may lose their protective mucous covering during isolation. For all these reasons, an enzyme in an isolated system may be more or less sensitive to the action of an inhibitor than it is in vivo.

The difficulty of predicting the site of action of an enzyme inhibitor in vivo from its effects on purified enzymes is illustrated by the demonstration that the activity of Na,K-ATPase in vitro is depressed by small concentrations of $HgCl_2$ (Nechay and Saunders, 1978). The Hg^{++} affinity for the purified enzyme is, indeed, very high (Anner, Moosmayer, and Imesh, 1992), but not necessarily higher than for some other cell constituents. Because the metal reacts avidly and nonspecifically with many proteins, the finding on the inhibition of the enzyme in vitro may possess little toxicologic significance; it clearly cannot prove that this enzyme is the target site for Hg^{++} under physiologic conditions in vivo.

Actually, there is direct evidence in another system against the hypothesis that in the intact cell the ATPase represents the primary target of Hg^{++}. Thus, Ballatori and Boyer (1996) observed that a micromolar concentration of the metal abolishes regulatory volume decrease (see Chapter 8) in skate hepatocytes; such an effect is not produced if ATPase activity is abolished with the classical inhibitor ouabain (Ballatori and Boyer, 1992). It follows that mercury inhibition of regulatory cell shrinkage cannot be attributed to inhibition of ATPase.

Problem of Extrapolation

A fifth problem arising from use of isolated systems is that of the necessary extrapolation back to the intact animal. This, of course, represents a classical difficulty with all in-vitro experiments and already is implied in the previous subsection. The simpler the model chosen for study, the greater the uncertainties attaching to any such extrapolation. Thus, the question may be asked whether qualitatively and quantitatively a purified membrane fragment reacts with a given xenobiotic in the same manner as do cell membranes in the intact organism.

Need for extensive extrapolation can be reduced by employing more intact preparations than purified membrane fragments. For instance, the isolated perfused renal tubule may provide more useful answers to specific questions about normal solute transport across the epithelial cell barrier in either direction than can purified preparations of brush border or basolateral cell membranes.

In conclusion, there are a number of factors that need to be taken into account carefully before chosing an in-vitro model for the analysis of a given problem. Such considerations must include the intrinsic structural or other limitations of an in-vitro model that can interfere with its desired application. The problem is reviewed under separate headings below for some of the more frequently applied in-vitro techniques in this field. Clearly, not all model preparations are appropriate for the study of all aspects of membrane toxicology.

3-3 TISSUE SLICES

Thin slices of many tissues long have been used in biomedical research. The extensive work of the Warburg school on tissue respiration was largely based on such preparations. Emphasis in these experiments focused on intracellular metabolism rather than on membrane function. A major concern was to cut the slices thinly enough so that diffusion of gas, substrates, metabolic products, and inhibitors into and out of the tissue could be neglected as a factor limiting solute turnover and metabolism. The assumption was made (not always correctly, as is seen subsequently) that under these conditions rates of substrate uptake and end-product extrusion do not limit the rate of metabolism and the general function of slices.

The fact that slices always must be cut as thinly as possible, and in any case to less than 0.3 mm, unavoidably increases the proportion of damaged cells. In spite of the presence of cell damage at the cut surfaces, however, slices have yielded important information in the study of solute transport across cell membranes and its depression by a variety of inhibitors. Examples are the uptake of glutamate by brain slices (Stern et al., 1949), the accumulation of organic anions by renal cortical slices (Forster,

1948; Cross and Taggart, 1950), and the maintenance of normal Na and K gradients, also in kidney slices (Mudge, 1951).

The important problem that must be reemphasized here is that for the morphologic reasons discussed below, membrane transport of organic solutes as well as inorganic electrolytes in renal cortical slices possesses little or no relevance to processes at the brush border, and more specifically to tubular reabsorption. Thus, slices of the aglomerular kidney of *Lophius americanus* readily transport potassium (Foulkes and Forster, 1959); this cannot represent a process of reabsorption of filtered potassium. Generally, electrolyte exchange in cortex slices presumably reflects little more than the homeostatic control of composition and volume typical of all normal cells (see Chapter 8).

Use of slices from complex tissues raises special questions related to tissue architecture and the probable loss of directional attributes of solute movement. This fact long has been recognized but has tended to be ignored by some investigators, both in earlier papers and following more recent improvements in preparation and handling of slices (see review by Foulkes, 1996). Specifically in renal cortical slices, it is evident that continued normal salt and water reabsorption from the tubular lumina in metabolically active tissue, in the absence of any filling pressure, must unavoidably lead to collapse of the tubule. This prediction was fully confirmed histologically by comparison of the open tubules in flash-frozen kidney slices with the collapsed tubules characteristic of slices incubated under usual conditions for 15 minutes (Boyesen and Leyssac, 1965).

Only if normal reabsorptive activity is inhibited would one expect to find open lumina in slices. This is precisely the observation reported by Ruegg, Gandolfi, Brendel et al. (1987), although the significance of this predictable finding was not recognized at the time. These authors observed mostly collapsed tubules in control slices, in contrast with many expanded lumina after short exposure to 1 mM $HgCl_2$ or $K_2Cr_2O_7$, or anoxia; the effect of Hg^{2+} is presented in Fig. 3-1. One might predict that such high metal concentrations will severely depress tissue metabolism and reabsorption of salt and water; accordingly only the control tubules should be collapsed.

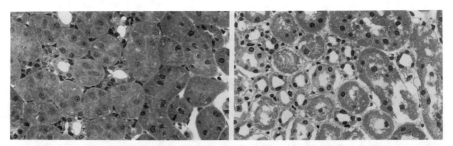

FIG. 3-1. Effect of Hg^{2+} on tubular collapse in cortical slices from rabbit kidneys. Control slices on left; on the right, slices exposed for 15 minutes to 1 mM Hg^{2+}. (From Ruegg, Gandolfi, Brendel et al., 1987; with permission.)

In summary, the collapse of tubules in metabolically active renal cortical slices from salt-conserving mammals can be predicted from two considerations: One is the ongoing and relatively large reabsorption of salt and water from luminal fluid; the second is the absence of filtration or any other force to replace the reabsorbed fluid. In addition, the length of the diffusion path between cells and suspending medium along even partially open lumina would greatly interfere with free solute exchange between medium and renal tubules at the brush border. As a result, apical processes of solute reabsorption out of, or secretion into the lumen of renal tubules, and their responses to toxic agents, cannot be studied adequately in mammalian slices.

Unlike in kidney slices from salt-conserving mammals with high filtered and reabsorbed loads of salt and water, in which tubules in slices can normally be expected to remain in a collapsed state, active luminal accumulation of an anionic dye like phenol red can be visualized readily in slices from a number of cold-blooded vertebrates (Forster, 1948). Presence of open tubular lumina in these preparations may be related to the fact that filtered sodium loads in these animals, and therefore the required rates of sodium reabsorption, generally are significantly smaller than in terrestrial mammals.

Not only membrane transport studies with various solutes but also metabolic measurements cannot rely always on renal cortex slices. As a consequence of the nonaccessibility of the luminal membrane in slices, an essential precondition for using them to study tissue metabolism is that uptake of substrates and inhibitors, as well as extrusion of products, can be assigned confidently to the basolateral cell membranes. For instance, exposure of renal cortical slices to $CdCl_2$ leads to rapid and presumably basolateral uptake of the metal; basolateral metal uptake frequently has been observed also in vivo (see Chapter 6). In contrast, Cd-metallothionein, according to results obtained in intact rabbits (Foulkes, 1978), is taken up only on the apical side of tubular epithelium; its ability to react with cortical slices therefore is limited. As a result, it clearly is inappropriate to compare the nephrotoxicity of the two Cd compounds with the slice technique. Independent of any improvements in their preparation and handling, renal cortical slices must be used only with full realization of their intrinsic limitations.

Although they clearly possess only limited usefulness for the study of apical solute movement, renal cortical slices have proven very useful in the analysis of basolateral transport processes. The best studied of these is the organic anion transport system, the focus of the classic work of Cross and Taggart (1950) on the mechanism of p-aminohippurate (PAH) secretion. The accumulation of PAH in these slices results from its active transport across the basolateral cell membranes (Foulkes and Miller, 1959). Accumulation of PAH is sensitive to a number of metabolic inhibitors and toxicants as illustrated in Table 3-1; competitive inhibition has also has been frequently reported.

The high slice/medium (S/M) concentration ratio (μmol/g slice divided by μmol/mL suspending medium) reached in control slices in Table 3-1 does, of course, reflect net uptake, not unidirectional influx into the tissue; moreover, after 60 minutes of incubation, the PAH concentration in slices (S) usually approaches a steady state. Slice-to-medium concentration ratios therefore generally should not be used to measure rates of uptake or to calculate fractional inhibition of transport following exposure

TABLE 3-1. *PAH accumulation by renal cortical slices and its inhibition*

	Slice/medium concentration ratio	Tissue water,[a] mL/g dry weight
Control	11.0	3.5
2:4 Dinitrophenol (50 μM)	1.2	—
Cyanide (0.5 mM)	1.3	—
HgCl$_2$ (1 mM)	1.4	5.7

[a]Representative values from the literature.
Mean values are shown for the concentration ratio (S/M), approaching steady state after 60 minutes of incubation.
Data from Cross and Taggart (1950).

to nephrotoxicants. Such rate calculations would be meaningful only as long as net uptake continues at the initial linear rate, indicating that S/M values have not approached steady state. In addition, a lowered steady-state S/M ratio need not indicate inhibition of active accumulation but could result of course from an increased rate of efflux.

Another source of uncertainty faced in attempts quantitatively to evaluate the activity of the organic anion transport system in renal slices is related to the fact that the S value of PAH usually is expressed on the basis of wet weight of slice, not the volume of cell water. Of course, this parameter is difficult to determine in renal tissue. The problem arises from the presence of unknown volumes of fluid in the tubules, containing unknown concentrations of PAH or extracellular markers; this is true especially under the influence of metabolic inhibitors interfering with the normal reabsorption of tubular fluid. It is not possible, therefore, to calculate in such a preparation intracellular solute concentrations in the normal fashion from total tissue contents of solute and water, with correction for the volume and composition of extracellular fluid. In addition, the sometimes large effects of metabolic inhibitors on tissue water content (see Table 3-1 and Chapter 8) further confound the significance of comparisons of PAH accumulation per unit wet weight. Accumulation can not be equated to intracellular PAH concentrations.

Further, even if the intracellular PAH content is known, its concentration in the cytoplasm cannot be computed in absence of a value for its volume of distribution in the cell (V_{PAH}); no measurements have been reported of V_{PAH}. To illustrate this problem mention may be made of the fact that cytoplasm (as opposed to total intracellular volume of the kidney) only constitutes about 13% of the total cortical volume of the rabbit kidney (Foulkes and Blanck, 1993). Even within the cytoplasm, PAH may not be evenly distributed, a possibility suggested by the finding of Miller, Stewart, and Pritchard (1993) that transcellular movement of organic anions in secreting epithelia results, at least in part, from vesicular movement.

Additional problems contribute to the difficulty of analyzing in slices the physiologic and toxicologic mechanisms of ion accumulation and its inhibition. For instance, the value of transmembrane electric potentials in slices seldom is known and may be significantly altered in the presence of toxic agents. As a result, steady-state distributions of ionic substrates can be depressed by an inhibitor that alters the potential without directly affecting the transporter.

The extent to which variations in membrane potential can represent a critical factor in the significance of S/M values is illustrated here with PAH. At a normal potential of -60 mV between cell and medium (cytoplasm negative), a PAH concentration in cytoplasm equal to that in the medium represents an electrochemical activity gradient of 10:1. Similarly, a luminal fluid compartment, if present and approximately 60 mV positive to cytoplasm, contains at equilibrium a ten times higher PAH concentration than cytoplasm. Clearly, the finding of S/M = 1.0 for PAH does not necessarily indicate abolition of active transport, as commonly suggested (e.g., Leibbrandt and Wolfgang, 1995). Effects of noncompetitive inhibitors of organic ion accumulation conceivably could be mediated by changes in transmembrane potential, rather than by a direct interaction with the anion transport system (see Chapter 5). In any case, S/M concentration ratios do not reflect actual electrochemical activity gradients of ionic substrates and therefore must be used cautiously in evaluating the extent of active transport and its inhibition. A reduced S/M ratio in renal cortical slices at steady state can provide at best only a qualitative measure of the effects of a toxicant on active transport.

The attribution of active transport in renal slices to the brush border membrane becomes especially dubious for substrates such as sugars and amino acids that react with both apical and basolateral cell membranes. Thus, Rosenberg, Blair, and Segal (1961) described the accumulation of natural amino acids in renal cortical slices and equated it to the physiologic process of proximal amino acid reabsorption from tubular urine. This identification is not compatible with the above conclusions about the role of the brush border in solute exchange in slices. It is thrown into further doubt by the subsequent finding that basolateral membranes in renal cortex of intact animals transport amino acids by carrier systems resembling those at the brush border (Silbernagl, Foulkes, and Deetjen, 1975). This resemblance not only refers to the substrate specificity of the carriers involved but also to their sensitivity to inhibition by, for instance, compounds of Hg (Foulkes, 1987).

Schafer and Barfuss (1980) further demonstrated with isolated, perfused proximal tubules of rabbit kidney (see section 3-6) that, like apical amino acid transport, basolateral amino acid uptake is an active process. The same conclusion was drawn from the steady state accumulation of the amino acid analogue cycloleucine in the renal cortex of intact rabbits (Foulkes and Blanck, 1991). Because accumulation of amino acids in vitro can be attributed to basolateral processes, no unique role of apical transport in slices should be assumed. This conclusion is independent of any a priori arguments based on tissue morphology and tubular collapse, and strengthens the interpretation that amino acid transport in slices possesses little relevance for the process of amino acid absorption across the brush border membranes.

Additional considerations strengthen the conclusion that renal cortical slices are unsuitable for the study of transport processes at the brush border. Foulkes and Miller (1959) calculated that a minimum of 80% of total PAH efflux from preloaded slices represents backflow across the basolateral membranes, rather than secretory movement across the brush border. Accumulation of L-lysine in kidney slices was separated from lysine reabsorption in the work of Ausiello, Segal, and Thier (1972). Finally, renal papillary slices (Lowenstein, Smith, and Segal, 1968) and isolated glomeruli

(MacKenzie and Scriver, 1971) accumulate amino acids even though these tissues are not involved in reabsorption of filtered amino acids. Substrate transport in these preparations, therefore, as well as in proximal tubules in slices in all probability reflects no more than the universal need of cells to take up essential metabolites. This presumably also accounts for amino acid accumulation in brain slices (Stern et al., 1949), although the neurotransmitter function of the amino acid also may play a role.

The relative inaccessibility of the brush border membrane in slices also limits their applicability to the study of metabolism and the actions of toxic agents. Use of slices for such studies is valid only to the extent that metabolic substrates or products, as well as inhibitors, can freely cross cell membranes. That this assumption is not always justified has been discussed here in relation to the metabolic effects of the low molecular weight protein Cd-metallothionein. As indicated, the protein does not react with basolateral cell membranes in the renal tubule and in vivo is only taken up at the brush border; uptake by slices is very slow. The metabolism of exogenous metallothionein and other protein substrates of the specific brush border carrier system, and their possible effects on cell function, therefore should not be studied in slices.

Failure to consider adequately the limitation on apical function in renal cortical slices still is encountered occasionally in the literature. For instance, Ruegg, Gandolfi, Brendel et al. (1987) reported on the differential patterns of injury to the proximal tubule in renal cortical slices following exposure to metal compounds. The relevance of these results to the intact kidney is doubtful, as the in-vitro findings reflect primarily metal uptake across the basolateral membranes. The suggestion that renal slices constitute a "valid in vitro system" for the kidney (Keith et al., 1995) obviously contradicts the interpretation of slice results offered in this chapter; the slice model appears valid only for limited functional purposes involving the basolateral cell membranes.

In spite of the presence of damaged cells, uncertainties about volumes of distribution and membrane potentials, and perhaps other factors, slices from tissues without the functional and structural complexity of renal cortex may yield results on membrane mechanisms and their intoxication that can be relatively unambiguously interpreted. In the case of renal cortical slices, however, the functional polarity of cells and the collapse of tubular lumina, with their implications for the accessibility of cell membranes to solutes from the bathing medium, raise legitimate questions about the general suitability of slices for metabolic studies and analysis of function at the brush border membrane. The extent to which slices from other tissues are limited similarly in their usefulness for quantitative measurement of function and of the effects of toxicants has not been considered fully.

3-4 RENAL TUBULE SUSPENSIONS

In attempts to design a more convenient and reproducible system than was offered by renal slices at the time, and in order to overcome some of the intrinsic disadvantages of slices, Burg and Orloff in 1962 utilized suspensions of short fragments of mostly proximal renal tubules for the study of renal transport mechanisms. The fragments were prepared by collagenase digestion of renal cortex and found frequent use in studies on renal solute transport. Nonperfused microdissected single-tubule fragments offer another approach to the study of tubular function. Very recently, Schafer et al. (1997) described a simplified method for isolating large numbers of defined nephron segments; the preparation involves incubation of manually disintegrated renal slices with collagenase and permits convenient isolation of fragments such as cortical collecting ducts, thick ascending limbs, and distal and connecting tubules.

An advantage of tubular fragment suspensions over the renal slice preparation is their ability to yield uptake rates measured over accurately timed short intervals; this makes possible determination of initial uptake velocities. Tubule suspensions continue to be applied successfully to study uptake of selected solutes, as illustrated by the findings of Groves et al. (1995) on renal tubular paraquat transport (see section 6-6).

Reliance on tubule fragments, however, poses its own set of potential problems. Among these remains especially the possibility that the method of preparation, usually involving enzymatic digestion with collagenase, may cause cell damage. Such damage has been observed by Mueller-Berger, Coppola et al. (1997) and Mueller-Berger, Nesterov, and Fromter (1997); the artifact was manifested by changes in the stoichiometry of sodium-bicarbonate cotransport, the response to acetazolamide, and the amiloride sensitivity of K conductance. The changes could be partially reversed by altering the conditions of incubation. A further problem is the uncertainty about the extent to which tubular lumina in fragments remain open, with consequently less restricted solute exchange at the brush border. Apical solute transport in fragments may contribute more to solute uptake and loss than it does in slices.

This conclusion is in full agreement with the observation by Murthy and Foulkes (1967) that the first-order rate constant of PAH efflux from slices (0.016 min^{-1}) approximates only a few percent of that from fragments (0.940 min^{-1}). The difference between rates of PAH efflux in the two preparations presumably reflects at least in part the ability of cells in fragments to secrete PAH in the normal direction across the apical cell wall. Predictably, the high efflux rate of PAH from fragments in these experiments was associated with lowered steady-state tissue-to-medium concentration ratios (16 for slices, 7 in fragments).

The fragment preparation is likely to offer significant although perhaps still restricted access to the brush border. As a result, it should be applied for transport studies only to solutes that do not react with the brushborder. Results on uptake of solutes like sugars or amino acids, normally transported at both cell membranes, are difficult to interpret.

3-5 EVERTED SACS OF SMALL INTESTINE

A variety of experimental models have been employed for studying the process of intestinal absorption and its inhibition by toxic agents. These preparations include isolated perfused segments of gut in vivo and in vitro, intestinal sacs, strips and rings, enteric cells in culture, and purified membrane vesicles from the intestinal epithelium. Each of these preparations exhibits, as usual, its own advantages and disadvantages; an overriding requirement always remains for continuous and adequate oxygenation. This led Fisher and Parson (1949) to the use of intestinal strips superfused on both sides with oxygenated solutions and prepared without any interval of anoxia.

A simpler preparation still minimizing anoxic damage to the intestinal mucosa was devised by Wilson and Wiseman (1954), who developed the everted sac technique for the study of absorptive processes in the small intestine. Such sacs first were prepared from hamsters but since have been obtained commonly from other small animal species, especially the rat. The objective of turning the tissue inside out is to provide maximal exposure of the mucosa to oxygenated medium. Transport then can be measured in the direction from mucosa to serosa, either as uptake into the tissue or as the equivalent of absorption, that is, transfer across the intestinal barrier into the serosal fluid contained inside the everted sac. To the extent that solutes like heavy metals are retained in the tissue, uptake from the mucosal fluid, of course, cannot be equated with absorption (see section 6-3).

Solute fluxes across the intestinal barrier require no mediation if they follow intercellular rather than transcellular pathways. The achievement of serosa/mucosa concentration gradients, however, as illustrated in Table 3-2, reflects the ability of this preparation actively to "absorb" solutes such as sugars and amino acids across cell membranes and along transcellular pathways. Direct evidence for the involvement of membranes is further provided by the finding of appropriate transporters in purified membrane vesicles. It follows that toxic inhibition of absorption cannot be attributed to indirect effects exerted, for example, on tight junctions.

Everted sacs have been relied on extensively in studies on solute absorption. Among these solutes, as shown in Table 3-2, are sugars and amino acids. Their movement against the concentration gradient from mucosa to serosa must involve active transport across cell membranes and is coupled to oxidative metabolism. It is inhibited by anoxia as well as by a variety of metabolic or competitive inhibitors. The uptake of

TABLE 3-2. *Solute transport in everted sacs of hamster small intestine*

| | | Concentration, *mM* | | |
| | | Final | | |
	Initial	Serosal	Mucosal	Serosa/mucosa
Glucose				
In O_2–CO_2	19	43	9	4.8
In N_2–CO_2	19	17	14	1.2
Methionine				
In O_2–CO_2	21	28	20	1.4
In N_2–CO_2	21	23	20	1.1

Initial concentrations were identical in serosal and mucosal fluids. Sacs were incubated for 60 minutes at 37°C.
Data from Wilson and Wiseman (1954).

a solute at the brush border represents the first step in its transcellular absorption (see section 6-3). The finding of various solute transporters in purified brush border membrane vesicles (see section 3-7) has provided additional support for the role of these membranes in intestinal solute absorption. For another transepithelial absorption process (renal tubular absorption), the primary role of the brush border has long been accepted. Kinetic evidence for this role in amino acid reabsorption and its inhibition was reported by Foulkes (1972).

The basolateral extrusion of an absorbed solute represents the second step in the process. Following this, extruded solutes in sacs must diffuse passively across the submucosal tissue, from cells into the serosal fluid. Unless significant amounts of solutes such as sugars or amino acids are retained in the submucosal tissue, this passive diffusion step should not appreciably reduce their appearance in the serosal fluid in the sac. Sacs thus may be accepted as useful models for the study of sugar and amino acid absorption and their inhibition by various toxicants.

This is not necessarily true of other solutes, and especially not of heavy metals. Nevertheless, attempts frequently have been reported to analyze the transfer of these metals across the intestinal barrier in sacs. The most important reason why the sac preparation cannot serve for study of transepithelial absorption of heavy metals is their well known and nonspecific binding to many tissue constituents. As a result, and unlike in the case of sugars and amino acids, the influence of submucosal tissue on the absorption of metals across sacs cannot be ignored. This conclusion is supported especially by the results of Endo et al. (1986), who removed the epithelial layer from everted sacs by exposure to 0.1 M EDTA and thereupon observed increased Hg uptake by the submucosal tissue. Apparently, the metal reacts with muscle and other tissue more readily than with the epithelial cells primarily responsible for metal transport.

In spite of the conclusion that subepithelial layers in sacs represent a major potential barrier for transmural metal movement, the first step of their absorption, their uptake by epithelial cells, can be studied conveniently in sacs. Results have been obtained with the sac technique on the first step of metal absorption, including metal–metal interaction at the brush border (Foulkes, 1985), on the effects of metabolic inhibitors, on specificity, saturability, the role of chelators, and other aspects of the uptake of heavy metals by jejunal epithelium (see section 6-3).

An additional difficulty associated with the use of sacs is posed by the doubts expressed about whether solute absorption follows identical pathways in the intact organ in situ and in everted sacs. In rat jejunum in vivo, metal absorption utilizes primarily transcellular pathways (see section 6-3). This conclusion is based on a variety of observations, including the kinetics of the process, its sensitivity to inhibitors, and its depression by the intracellular metal-trapping protein metallothionein, as reported by Foulkes and McMullen (1986) in rats. Mention must be made here of the fact that Lind and Glynn (1997) could not confirm in mice the suggested trapping role of metallothionein in the control of cadmium absorption. Even though the dose of $CdCl_2$ per animal in those experiments was small, the concentration in the fluid orally administered can be calculated to have been as high as about 10 mM; the possibility of localized intestinal toxicity, therefore, cannot be neglected. In any

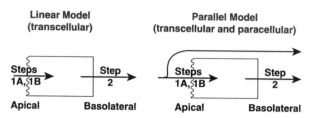

FIG. 3-2. Kinetic models of intestinal absorption.

case, cellular uptake at the brush border does appear to constitute the initiating step in metal absorption under physiologic conditions.

The problem now arises that the linear kinetic model proposed in Fig. 3-2, which fits the transfer of metals from lumen into portal blood in the living rat, has been assumed implicitly in many studies to describe also the process of metal absorption in sacs; this does not appear to be correct. Instead, a possibly significant contribution to metal absorption in sacs made by diffusion pathways between cells (paracellular absorption) may require the parallel model also shown in Fig. 3-2.

Figure 3-2 schematically depicts an intestinal epithelial cell, with either an exclusively transcellular route of absorption from the brush border (linear model, at left) or, in contrast, with absorption following in part a paracellular pathway (parallel model, at right). Cellular uptake in both cases is believed to result from a binding step 1A followed by an internalization step 1B (see section 6-3). Note that step 2, defined as the basolateral extrusion of an absorbed solute, may account in the parallel model for only a small contribution to total absorption.

Choice of an inappropriate model to describe absorption in everted sacs, of course, leads to erroneous interpretation of results obtained. This fact is apparent in the attempts to explain acute inhibition of Cd transport in everted sacs by millimolar concentrations of EDTA. On short exposure at low concentrations, this chelator removes only extracellular and membrane-bound Cd from the intestinal mucosa (see section 6-3); it also prevents the metal from reacting with the cell membrane in the first place. The report that at higher concentrations EDTA decreases cellular uptake of Cd in everted sacs and increases its transmural movement (Kojima and Kiyozumi, 1974) cannot be explained readily on the basis of the linear model. Instead, a more likely explanation is provided by the parallel model. According to this view, EDTA depresses cellular Cd uptake but opens up parallel and presumably intercellular pathways for movement of the metal across the cell barrier. Opening of Ca-gated solute channels in the intestinal mucosa by EDTA has been documented previously, as in the work of Tidball (1964) on phenol red uptake.

The possibility that the handling of sacs during preparation may lead to artifacts that permit greater movement of solutes through aqueous pores, between cells, or into cells was tested in rats by comparison of urea fluxes out of mucosal fluid in vivo and in everted sacs (Foulkes and Bergman, 1993). Fluxes of urea were normalized by the simultansouly measured efflux of ethanol, a measure at a given concentration gradient and temperature of the size of the absorbing area (Foulkes, Mort, and Buncher, 1991).

The normalized urea flux thus provides in arbitrary units an estimate of the density of aqueous pores per unit membrane area. As detailed in Table 6-2, a four-fold increase in the density of such pores was found to be associated with the preparation of everted sacs of jejunum from adult rats.

The flux of urea, as stated, provides a measure for the presence of polar diffusion pathways, including pathways between cells (paracellular pathways); heavy metals presumably can diffuse through such paracellular pathways. To the extent that aqueous pores represent passage between cells, the parallel model is more appropriate than the linear model A for describing metal absorption in sacs. This possibility was not adequately considered in some reports on metal absorption in everted sacs (e.g., Kojima and Kiyozumi, 1974; Ohta, De Angelis, and Cherian, 1989); as a result, the significance of the reported findings remains open for reinterpretation.

The overall conclusion drawn from all these observations is that the everted intestinal sac cannot serve as a suitable physiologic model for quantitatively analyzing the overall process of absorption, at least of heavy metals, from the intestinal lumen into blood. Just as in renal cortical slices, which can serve as a model for solute exchange on only one side (basolateral) of tubule cells, so in everted sacs it is primarily the apical membrane of mucosal epithelium that can provide quantitative information on solute transport.

Although for the reasons just described the physiologic and toxicologic significance of solute transport in everted sacs and its inhibition by toxicants is not always self-evident, a number of investigators have used the preparation to study the action of toxic xenobiotics on intestinal absorption. Such use, as just indicated, is appropriate for the analysis of the first step in transcellular absorption pathways, the uptake of the test solute by the mucosal cells (see Fig. 3-2). The variable influence of submucosal tissue on movement of solutes, and especially of heavy metals, makes sac preparations less applicable to the study of basolateral processes. This may leave open the question of to what extent toxic effects on solute movement across mucosal epithelium may be exerted at the basolateral cell membranes.

An instance of multiple sites of toxicant action on epithelial transfer is that of mercury on amino acid absorption in the renal tubule (see Table 7-1). The metal depresses function at the apical membrane, as reflected in reduced amino acid removal from the tubular lumen. At the same time, the transcellular passage time of the amino acid is prolonged greatly (Foulkes, 1987), and its steady-state concentration in the cells is raised. On the basis of the linear model of transepithelial transport (see Fig. 3-2), these results are best interpreted in terms of inhibition of both apical amino acid uptake and basolateral extrusion. In other words, mercury and other metals (Foulkes and Blanck, 1991) under these conditions depress amino acid transfer at more than one site; this does not necessarily imply that the metal-sensitive transport systems on the two opposing membranes must be identical. Tables 5-1 and 7-1 provide additional details of multiple cellular sites of acute action of mercury.

In conclusion, study of the second step in transcellular absorption of solutes in the intestine, their extrusion across the basolateral cell membrane, and its inhibition by toxicants is rendered difficult by the presence of extensive submucosal tissue. Especially in the case of solutes like heavy metals, likely to react with this tissue, the extrusion mechanism, therefore, cannot readily be studied accurately in everted sacs. It is only for the study of apical processes that the sac preparation provides a suitable model for quantitative study. Even for that limited purpose, however, evidence has been obtained for apparent artifacts of preparation.

3-6 PERFUSED INTACT TISSUES OR TISSUE FRAGMENTS

Membrane function and its inhibition also have been successfully investigated in tissues artificially perfused in situ or in vitro; a classic example is the use of intact perfused kidneys. Although it is technically no simple task in this preparation to control such variables as flow resistance of the vascular system, to prevent swelling, and to maintain near normal fractional Na and water reabsorption for long time intervals, successful preparations have been available for many years, capable of maintaining high glomerular filtration rates and excreting relatively concentrated urine.

Perfused organs possess many advantages for the study of tissue metabolism and solute transport across membranes; three of these advantages are cited here. First, of course, the intact organ presents fewer difficulties of extrapolation back to the whole animal than do more simplified preparations. Second, it permits study of a much wider range of xenobiotic concentrations than can be achieved in vivo. Third, provided the oncotic pressure of the perfusates remains normal, blood can be replaced by synthetic solutions of accurately defined composition. This possesses a number of obvious advantages, such as better control of speciation of toxicants like heavy metals or the selection of preferred metabolic substrates.

In general, technical difficulties attaching to vascular organ perfusion with blood or synthetic perfusates largely have been overcome, and the approach has been successfully applied to many organs and regions, including kidneys, liver, mesentery, pancreas, heart, and lungs. Care must be taken during blood perfusion that presence of anticoagulants does not interfere with normal handling of certain solutes such as the heavy metals. Heparin or citrate, for instance, are likely to alter diffusibility and filterability of heavy metals (Foulkes, 1977; Vander et al., 1977).

Luminal perfusion of organs like renal tubules or segments of intestine also has yielded extensive insights into membrane function. For instance, microperfusion of single tubules in vivo, or of isolated single-tubule segments in vitro, have been applied to many studies in renal toxicology. The technique has been applied, for instance, to study reactions of Cd (Felley-Bosco and Diezy, 1987) and Hg (Zalups and Barfuss, 1990) with brush border membranes. Solute movement between the tubular lumen (apical side) of isolated segments and the bathing medium (basolateral side) can be followed readily, as can direct effects of toxicants on both apical and basolateral membranes.

In gastrointestinal toxicology, segments of intestine perfused through both the lumen and the vascular system have found use, for instance, in the analysis of heavy metal absorption (Smith et al., 1978). The results obtained in this preparation, however, do not agree fully with what has been observed in the intact animal. In particular, the transmural absorption of Cd proceeds much faster in the doubly perfused intestine than it does in the intact animal. This perhaps can be explained by such artifacts in the doubly perfused organ as increased intercellular instead of transcellular absorption pathways, as described in the previous section for everted sacs.

3-7 MEMBRANE PATCHES AND VESICLES

Membrane patches and purified membrane vesicles constitute some of the most simplified experimental models for the study of membrane function and its inhibition by toxic substances. They make it possible to study independently the effects of altering the composition of the media bathing the "inner" and "outer" (i.e., intracellular and extracellular) aspects of the membrane. Similarly, action of toxic agents can be compared in these preparations on the opposing sides of the membrane. It is possible with the patch technique to focus on, for example, a single membrane channel for an ionic solute.

A further refinement reconstitutes synthetic membranes by incorporating isolated constituents such as ion channels into synthetic lipid layers. The advantages of reducing complexity by studying function in such relatively simple models are obvious. So are some of the disadvantages of these preparations, including especially the need for far-ranging extrapolations to the intact organism. In addition, one must consider the possible loss of membrane constituents or changes in membrane function during isolation. Morris and Horn (1991), for instance, concluded that the patch-clamp technique hypersensitizes channels to stretch. Another limitation of these highly simplified preparations is illustrated by the fact that physiologic control of the activity of a solute channel in vesicles is likely to be lost if its gating behavior depends, for instance, on phosphorylation. In the absence of normal control mechanisms, the isolated preparation tells us how channels can behave, not necessarily how they do behave in a more physiologic setting.

Nevertheless, such "simplified" preparations have been applied successfully to many problems in membrane toxicology. The patch electrode technique developed by Hamill et al. (1981) possesses the great advantage that it makes possible the functional analysis of a presumably intact structure exposed to essentially physiologic bathing media. It has been used to demonstrate, for instance, responses of single-ion channels to cell depolarization and to stretch, or to some of the toxicants listed in Table 5-3.

Purified membrane vesicles capable of carrying out some of the active transport processes of intact cells repeatedly have been prepared by many investigators (see, e.g., Murer and Kinne, 1980) and have been applied extensively to study effects of toxicants on membrane function. Active uptake is measured over short time intervals, during which the transport substrate may transiently achieve an intravesicular peak concentration in excess of the equilibrium level (the so-called overshoot phenomenon). Energy for this active accumulation may be provided by coupling to some ion gradient across the membrane. Subsequently, dissipation of this gradient returns the transmembrane distribution of the solute to equilibrium. By experimentally altering such ion gradients it becomes possible to distinguish between effects of toxic agents exerted either directly on membrane transporters or indirectly by abolishing the ion gradient (see Chapter 5).

Several laboratories have employed vesicle preparations to study membrane transport of metals and their toxic effects on membrane function. Metal accumulation in vesicles generally is not affected by osmotic shrinkage or swelling in anisotonic solutions. In the absence of intravesicular metal ligands this finding presumably reflects complete binding of the accumulated metals to membrane constituents. In the intact cell, in contrast, accumulated metals are also likely to react with a wide variety of cytoplasmic and other metal ligands. Further, comparison of function in vivo and in vitro, as emphasized earlier, may be vitiated, especially if the membranes are exposed to different species of a toxicant under these conditions. An obvious example here is the study of vesicles reacting with ionized metals, as contrasted with intact tissue in situ exposed to various metal–ligand complexes. All these considerations again reinforce the view that highly purified and simplified preparations like membrane vesicles may tell us how membranes can react but not necessarily how they do react in the intact organism.

Among the results achieved in the study of metal transport by vesicles are those reviewed by Cousins (1985) on the uptake of Zn. The mechanism of renal Pb uptake was studied by Victery, Miller, and Fowler (1984) in vesicles from the renal cortex. Bevan and Foulkes (1989) found that the reaction of Cd with intestinal brush border membrane vesicles resembles in some regards that described for intact tissue. In both preparations, metal binds to the membrane in the cold, but internalization of the bound metal is a temperature-dependent process as described in section 6-3.

The internal localization of some of the binding sites can be inferred from the fact that, for instance, only portion of bound cadmium can be removed by an extravesicular chelator like EDTA. It is only following destruction of membrane integrity that the metal becomes fully accessible to the chelator and consequently can be extracted completely (Bevan and Foulkes, 1989). Similarly, intravesicular binding of Cd is suggested by the observation that glucose uptake by vesicles from the mucosal brush border is inhibited by preincubation with Cd but this inhibition could not be reversed by EDTA. Bevan et al. (1989) studied the active accumulation of alanine by renal brush border membrane vesicles from the winter flounder; this transport was depressed by Cd only after a significant initiation delay.

Such observations can be explained by assuming that the target site of the metal here is deeply buried in the membrane or even is localized on the inner (cytoplasmic) aspect of the membrane. The initiation delay for the toxic action of metals in vesicles, similar to that previously demonstrated in the intact animal (see section 5-4), then may reflect the time required for critical concentrations of the toxicant to reach its site of action in the membrane. Possible alternative interpretations for the initiation delay might invoke an indirect effect of the metal on membrane function or a gradually developing and probably nonspecific lesion such as would be caused by, for example, peroxidation of membrane lipids. Such possible modes of action are further considered in Chapter 5.

Another illustration of the usefulness of purified membrane vesicles for the study of transport of toxic agents is the finding by Selenke and Foulkes (1981) that Cd-metallothionein reacts with renal brush border vesicles in a manner similar to its reaction with the brush border of the proximal tubule during reabsorption in vivo (see Chapter 6). In both systems, two parallel reactions of Cd-metallothionein with brush border membranes were observed. Myoglobin inhibits Cd-metallothionein reabsorption in vivo and Cd-metallothionein binding to brush border membrane in vitro. Finally, reabsorption of Cd-metallothionein is depressed in rabbits chronically treated with Cd, and brush border vesicles isolated from such animals show a decreased ability to bind Cd-metallothionein.

In summary, many important findings relating to membrane toxicology have been made with the vesicle technique. Nevertheless, it is difficult always to exclude the possibility that isolation and purification of vesicles may be accompanied by artifacts such as alteration or loss of membrane constituents. This fact complicates the extrapolation of findings made with vesicles to the intact organism.

3-8 CONCLUSIONS

The study of toxic action in simplified preparations may provide insight into how a toxicant can inhibit function of certain membranes but not necessarily how and where it does act in the intact organism. This is especially true of effects exerted indirectly on the membranes (see Chapter 5). Obviously, extrapolation from in-vitro experiments on membrane function to the intact organism requires great care. In addition, the intrinsic limitations especially of simplified experimental preparations must be appreciated fully, lest questions be asked that the proposed experimental model is a priori unable to answer. With such precautions, the in-vitro approaches to the study of membrane toxicology undoubtedly will continue to yield important information.

REFERENCES

Anner, B.M., M. Moosmayer, and E. Imesh. 1992. Mercury blocks Na-K-ATPase by a ligand-dependent and reversible mechanisms. *Am. J. Physiol.* 262:F830–F836.

Ausiello, D., S. Segal, and S.O. Thier. 1972. Cellular accumulation of L-lysine in rat kidney cortex in vivo. *Am. J. Physiol.* 222:1473–1478.

Ballatori, N., and J.L. Boyer. 1992. Taurine transport in skate hepatocytes: II. Volume activation, energy and sulfhydryl dependence. *Am. J. Physiol.* 262:G451–G460.

Ballatori, N., and J.L. Boyer. 1996. Disruption of cell volume regulation by mercuric chloride is mediated by an increase in sodium permeability and inhibition of an osmolyte channel in skate hepatocytes. *Toxicol. Appl. Pharmacol.* 140:404–410.

Bevan, C., and E.C. Foulkes. 1989. Interaction of cadmium with brush border membrane vesicles from the rat small intestine. *Toxicology* 54:297–309.

Bevan, C., E. Kinne-Saffran, E.C. Foulkes, and R.K.H. Kinne. 1989. Cadmium inhibition of L-alanine transport into renal brush border membrane vesicles isolated from the winter flounder. *Toxicol. Appl. Pharmacol.* 101:461–469.

Boyesen, E., and P. Leyssac. 1965. The kidney cortex slice as a model for sodium transport in vivo. *Acta Physiol. Scand.* 65:20–32.

Burg, M.B., and J. Orloff. 1962. Oxygen consumption and active transport in separated renal tubules. *Am. J. Physiol.* 203:327–330.

Cousins, R.J. 1985. Absorption, transport and hepatic metabolism of copper and zinc: Special reference to metallothionein and ceruloplasmin. *Physiol. Rev.* 65:238–309.

Cross, R.J., and J.V. Taggart. 1950. Accumulation of PAH by rabbit kidney slices. *Am. J. Physiol.* 161:181–190.

Endo, T., S. Nakaya, R. Kimura, and T. Murata. 1986. Gastrointestinal absorption of inorganic mercury compounds in vitro. *Toxicol. Appl. Pharmacol.* 83:187–196.

Felley-Bosco, E., and J. Diezy. 1987. Fate of cadmium in renal tubules: A microinjection study. *Toxicol. Appl. Pharmacol.* 91:204–211.

Fisher, R.B., and D.S. Parson. 1949. A preparation of surviving rat small intestine for the study of absorption. *J. Physiol.* 110:36–46.

Forster, R.P. 1948. Use of thin kidney slices and isolated renal tubules for direct study of cellular transport kinetics. *Science* 108:65–67.

Foulkes, E.C. 1972. Cellular localization of amino acid carriers in renal tubules. *Proc. Soc. Exp. Biol. Med.* 139:1032–1033.

Foulkes, E.C. 1977. Mechanism of renal excretion of environmental agents. In *Handbook of physiology, section 9: reactions to environmental agents*, ed. D.H.K. Lee, 495–502. Washington: American Physiological Society.

Foulkes, E.C. 1978. Renal tubular transport of cadmium metallothionein. *Toxicol. Appl. Pharmacol.* 45:505–512.

Foulkes, E.C. 1985. Interactions between metals in rat jejunum: Implications on the nature of cadmium uptake. *Toxicology* 37:117–125.

Foulkes, E.C. 1987. Role of basolateral cell membranes in organic solute reabsorption in rabbit kidneys. *Am. J. Physiol.* 252:F1042–F1047.

Foulkes, E.C. 1996. Slices and sacs: Limitations on metabolic and functional studies in kidney cortex and intestine. *Proc. Soc. Exp. Biol. Med.* 211:155–162.

Foulkes, E.C., and D. Bergman. 1993. Inorganic Hg absorption in mature and immature rat jejunum: Transcellular and intercellular pathways in vivo and in everted sacs. *Toxicol. Appl. Pharmacol.* 120:89–95.

Foulkes, E.C., and S. Blanck. 1991. Cadmium inhibition of basolateral solute fluxes in rabbit renal tubules and the nature of cycloleucine uptake. *Toxicol. Appl. Pharmacol.* 108:150–156.

Foulkes, E.C., and S. Blanck. 1993. Volume of renal cortical cytoplasm in rabbits and basolateral transport gradients of cycloleucine and PAH. *Proc. Soc. Exptl. Biol. & Med.* 202:302–306.

Foulkes, E.C., and R.P. Forster. 1959. Potassium transport by kidney slices of Lophius americanus. *Bull. Mt. Desert Island Biol. Lab.* 4:44.

Foulkes, E.C., and D.M. McMullen. 1986. Endogenous metallothionein as determinant of intestinal cadmium absorption: A reevaluation. *Toxicology* 38:285–291.

Foulkes, E.C., and B.F. Miller. 1959. Steps in PAH transport by kidney slices. *Am. J. Physiol.* 196:86–92.

Foulkes, E.C., T. Mort, and R. Buncher. 1991. Intestinal cadmium permeability in mature and immature rats. *Proc. Soc. Exp. Biol. Med.* 197:477–481.

Groves, C.E., M.N. Morales, A.J. Gandolfi, W.W. Dantzler, and S.H. Wright. 1995. Peritubular paraquat transport in isolated renal proximal tubules. *J. Pharmacol. Exp. Ther.* 275:926–932.

Hamill, D.P., A. Marty, E. Neher, B. Sakmann, and F.J. Sigworth. 1981. Improved patch-clamp technique for high resolution current recording from cells and cell-free membrane patches. *Pflugers Arch.* 391:85–100.

Keith, R.L., S.J. McGuiness, A.J. Gandolfi, T.P. Lowe, Q. Chen, and Q. Fernando. 1995. Interaction of metals during their uptake and accumulation in rabbit renal cortical slices. *Environ. Health Perspect.* 103 (suppl. 1):77–80.

Kojima, S., and M. Kiyozumi. 1974. Studies on poisonous metals I: Transfer of cadmium chloride across rat small intestine in vitro and effect of chelating agents on this tranfer [Japanese]. *Yakugaku Zasshi* 94:695–701.

Leibbrandt, M.E.I., and G.H.I. Wolfgang. 1995. Differential toxicity of cisplatin, carboplatin, and CL-973 correlates with cellular platinum levels in rat renal cortical slices. *Toxicol. Appl. Pharmacol.* 132:245–252.

Lind, Y., and A.W. Glynn. 1997. The involvement of metallothionein in the intestinal absorption of cadmium in mice. *Toxicol. Lett.* 91:179–187.

Liu, J., Y. Liu, and C.D. Klaassen. 1994. Nephrotoxicity of $CdCl_2$ and Cd-metallothionein in cultured rat kidney proximal tubules and LLC-PK$_1$ cells. *Toxicol. Appl. Pharmacol.* 128:264–270.

Lowenstein, L.M., I. Smith, and S. Segal. 1968. Amino acid transport in the rat renal papilla. *Biochim. Biophys. Acta* 150:73–81.

Mackenzie, S., and C.R. Scriver. 1971. Transport of L-proline and α-aminoisobutyric acid in the isolated rat kidney glomerulus. *Biochim. Biophys. Acta* 241:725–736.

Miller, D.S., D.E. Stewart, and J.B. Pritchard. 1993. Intracellular compartmentation of organic anions within renal cells. *Am. J. Physiol.* 264:R882–R890.

Morris, C.E., and R. Horn. 1991. Failure to elicit neuronal macroscopic mechanosensitive currents anticipated by single-channel studies. *Science* 251:1246–1249.

Mudge, G.H. 1951. Electrolyte and water metabolism of rabbit kidney slices: Effects of metabolic inhibitors. *Am. J. Physiol.* 167:206–223.

Mueller-Berger, S., S. Coppola, I. Samarzija, G. Seki, and E. Fromter. 1997. Partial recovery of in vivo function by improved incubation conditions of isolated renal proximal tubule: I. Change of amiloride-inhibitable K$^+$ conductance. *Pflugers Arch.* 434:372–382.

Mueller-Berger, S., V.V. Nesterov, and E. Fromter. 1997. Partial recovery of in vivo function by improved incubation conditions of isolated renal proximal tubule: II. Change in Na-HCO$_3$ cotransport stoichiometry and of response to acetazolamide. *Pflugers Arch.* 434:383–391.

Murer, H., and R.K.H. Kinne. 1980. The use of isolated membrane vesicles to study epithelial transport processes. *J. Membr. Biol.* 55:81–95.

Murthy, L., and E.C. Foulkes. 1967. Solute movement across luminal cell membranes in rabbit kidney tubules. *Nature* 213:180–181.

Nechay, B.R., and J.P. Saunders. 1978. Inhibitory characteristics of cadmium, lead and mercury in human sodium and potassium dependent adenosine triphosphatase preparations. *J. Environ. Pathol. Toxicol.* 2:283–290.

Ohta, H., M.V. De Angelis, and M.G. Cherian. 1989. Uptake of cadmium and metallothionein by rat everted intestinal sacs. *Toxicol. Appl. Pharmacol.* 101:62–69.

Rosenberg, L.E., A. Blair., and S. Segal. 1961. Transport of amino acids by slices of renal cortex. *Biochim. Biophys. Acta* 54:479–488.

Ruegg, C.E., A.J. Gandolfi, K. Brendel, and R.B. Nagle. 1987. Differential patterns of injury to the proximal tubule of renal cortical slices following in vitro exposure to $HgCl_2$, $K_2Cr_2O_7$ or anoxic conditions. *Toxicol. Appl. Pharmacol.* 90:261–273.

Ruegg, C.E., A.J. Gandolfi, R.B. Nagle, C.L. Krumdieck, and K. Brendel. 1987. Preparation of positional renal slices for study of cell-specific toxicity. *J. Pharmacol. Methods* 17:111–123.

Schafer, J.A., and D.L. Barfuss. 1980. Membrane mechanisms for transepithelial aminoacid absorption and secretion. *Am. J. Physiol.* 238:F335–F341.

Schafer, J.A., M.L. Watkins, L. Li, P. Herter, S. Haxelmans, and E. Schlatter. 1997. A simplified method for isolation of large numbers of defined nephron segments. *Am. J. Physiol.* 273:F650–F657.

Schultz, S.G. 1996. Homeostasis, Humpty Dumpty, and integrative biology. *News Physiol. Sci.* 11:238–246.

Selenke, W., and E.C. Foulkes. 1981. The binding of cadmium metallothionein to isolated renal brush border membranes. *Proc. Soc. Exp. Biol. Med.* 167:40–44.

Silbernagl, S., E.C. Foulkes, and P. Deetjen. 1975. Renal transport of amino acids. *Rev. Physiol. Biochem. Pharmacol.* 74:105–167.

Smith, K.T., R.J. Cousins, B.L. Silbon, M.L. Failla. 1978. Zinc absorption and metabolism by isolated vascularly perfused rat intestine. *J. Nutr.* 108:1849–1857.

Stern, J.R., L.V. Eggleston, R. Hems, and H.A. Krebs. 1949. Accumulation of glutamic acid in isolated brain tissue. *Biochem. J.* 44:410–418.

Tidball, C.S. 1964. Magnesium and calcium as regulators of intestinal permeability. *Am. J. Physiol.* 206:243–246.

Vander, A.J., D.L. Taylor, K. Kalitis, D.R. Mouw, and W. Victery. 1977. Renal handling of lead in dogs: Clearance studies. *Am. J. Physiol.* 233:F532–F538.

Victery, W., C.R. Miller, and B.A. Fowler. 1984. Lead accumulation by rat renal brush border membrane vesicles. *J. Pharmacol. Exp. Ther.* 231:589–596.

Wilson, T.H., and G. Wiseman. 1954. The use of everted sacs of small intestine for the study of the transference of substances from mucosal to serosal surfaces. *J. Physiol.* 123:116–125.

Zalups, R.K., and D. Barfuss. 1990. Accumulation of inorganic mercury along the renal proximal tubule of the rabbit. *Toxicol. Appl. Pharmacol.* 106:245–253.

4

Constituents, Properties, and Functions of Cell Membranes

4-1 INTRODUCTION

The toxicologic significance of biologic membranes is considered in Chapter 2. This significance stems from the general dependence of the toxicokinetics of nonlipophilic xenobiotics on the properties of cell membranes, and the sensitivity of individual membrane functions to the direct or indirect toxic action of many xenobiotics. A detailed discussion of the underlying structure and function of cell membranes lies well beyond the scope of the present work. It would have to cover the chemistry of membrane constituents and their placement in the membrane structure, active and facilitated solute transport processes, the nature and control of membrane channels for solutes and water, action of enzymes at the membrane, receptor pharmacology, the role of membranes in cell recognition, rejection, communication and adhesion, and other membrane properties.

The ability of membranes to carry out complex functions implies a complex structure and composition. In order to understand how membranes perform their functions, and by what mechanisms toxic agents can interfere with these functions, it is essential to develop an appreciation of the nature of membranes and of their metabolism. It is the aim of this chapter to provide the necessary background information for this purpose. More extensive information on these topics, however, is available readily in numerous authoritative books and comprehensive reviews. A useful recent publication is that of Byrne and Schultz (1997).

Existence of a limiting structure separating cells from their surroundings was implicit in the discovery of cells themselves. Even though it was only much later that

41

plasma membranes surrounding animal cells actually could be observed under the microscope, implicit recognition of a semipermeable barrier separating cytoplasm from extracellular fluid had emerged by 1773 (see Chapter 1). The existence of such a membrane was deduced from the reaction of erythrocytes (although not of most other cells, as discussed in Chapter 8) to what we now call an osmotic gradient; over a wide range of extracellular osmolarities these cells behave like perfect osmometers. The lipid nature of the separating membrane subsequently was suggested by the selective permeability of cells to lipid-soluble compounds and by the cytolytic effect of detergents. Erythrocytes, for instance, are readily hemolyzed by low concentrations of compounds such as n-butanol and saponin.

The classic model of membrane structure was formulated by Davson and Danielli (1952). It basically consists of a bilayer of lipoproteins, whose polar groups project into the aqueous phase. As is developed further in section 4-3, the basic bilayer structure does not remain static but possesses significant fluidity and may undergo rapid turnover in response to a variety of environmental, physiologic, or toxicologic factors. This so-called membrane traffic is referred to in section 2-6 and forms the topic of section 4-3.

A significant number of specific peptides and proteins, complex carbohydrates, and phospholipids are supported in and on this lipoprotein matrix; they are largely responsible for the many specialized functions of the membranes. These functions, as already indicated, are in many instances sensitive to toxic agents. A typical example, further discussed in section 4-4, is the interference by Pb with cellular adhesion in the central nervous system; the heavy metal here reacts with the neural cell–adhesion molecule, a cell-surface glycoprotein (see review by Reuhl and Dey, 1996). The reactions of surface molecules with toxicants are discussed in greater detail especially in Chapters 5 and 6.

Although the various molecules on the cell surface may be anchored in the membrane, usually through hydrophobic domains, their hydrophilic domains can react with solutes in the aqueous phase, inside or outside the cell. This has been established, for instance, for the so-called ectoenzymes (see section 4-4), which catalyze extracellular reactions near the membrane. In general terms, the specialized membrane constituents are intimately involved in interactions between cells and their environment and in the control of cell composition and volume.

In addition to specific macromolecules anchored in the membranes, complex but nonspecific molecules (mucins) also are found associated with the outer surface of cell membranes; some of these compounds may be anchored in the membrane (Khatri, Forstner, and Forstner, 1997). Mucins are composed largely of glycosaminoglycans and coat the surface of many cells (see, e.g., Nimmerfall and Rosenthaler, 1980). They are catalytically inactive but protect the membrane against noxious extracellular agents, such as high acidity in the stomach. Binding to mucins can lower the concentration of reactive heavy metals at the membrane. At the same time, presence of mucins also increases the thickness of so-called unstirred water layers next to the membrane, a fact that strongly can influence the kinetics of solute exchange across the membranes.

Even though such mucins serve primarily as passive protective coating of cells, they themselves may become targets of toxic action. An example is the mucous layer on the surface of bladder epithelium, whose antiadherence effect is inactivated by the complete urinary carcinogen N-methyl-N-nitrosourea (Bodenstab, Kaufman, and Parsons, 1983). This observation further emphasizes the broad spectrum of potential actions of toxicants at the cell membrane.

4-2 REACTIONS AT THE MEMBRANE

As already mentioned, a general, although not necessarily exhaustive, list of specialized membrane-based processes includes reaction of many hormones and drugs with their specific receptors, and the subsequent transmission of the message received across the membrane (see also section 4-5); function of membrane constituents as specific antigens; participation in intercellular communication, as in the case of gap junction proteins such as the connexins; a variety of enzyme-catalyzed reactions; control of the transmembrane movement of solutes such as substrates, waste products, and others required for the maintenance of constant cell composition and volume (see Chapter 8); and the establishment of the electrochemical activity gradients across the membrane basic for the function of excitable and other cells. Most of these processes, of course, require ready accessibility of membrane sites to extracellular solutes. Within the present context, this fact implies that these sites on cell membranes also are exposed readily to toxic xenobiotics.

Presence in or on the membrane of complex proteins to mediate all these and other specific functions permits the prediction that general protein reagents are likely to react and interfere relatively nonspecifically with membrane function. No particular toxicologic significance therefore can be attached, for instance, to the observation that mercuric mercury inhibits membrane Na,K-ATPase in vitro (Nechay and Saunders, 1978). As further considered in section 5-4, and unless it can be shown that the affinity of this enzyme for Hg^{2+} clearly exceeds that of other proteins, such a finding can yield little information on the nature of the primary target of this toxicant in vivo.

Although some functional membrane lesions result directly from interactions between toxic agents and critical molecules at the membrane, others may represent an indirect consequence of reactions elsewhere on the membrane or in the cell, or even in other tissues. An example of such action at a distance is the depression of renal sodium reabsorption following damage to the adrenal cortex and consequent reduced aldosterone secretion. Both direct and indirect actions may affect specifically a particular function or interfere nonspecifically with general membrane integrity. An example of a membrane reaction likely in this manner to lead to nonspecific functional lesions is the peroxidation of membrane lipids in the presence of metals and other reagents. As further discussed in Chapter 5, it is difficult at times clearly to distinguish between direct and indirect actions of toxic agents on membrane function.

Cell membranes generally carry a fixed net negative electric charge, associated primarily with anionic groups on proteins, sialic acid, sulfonic acid groups in complex carbohydrates including glycoproteins, and the phosphate groups of phospholipids such as phosphatidylinositol and phosphatidylserine. Charge distribution and transmembrane potentials determine many reactions of cells with their environments. This is seen, for instance, in the voltage gating of selected solute channels (see section 4-6). Charge distribution on the membrane is greatly altered following, for example, removal of sialic acid residues with neuraminidase or the neutralization of fixed anionic charges by cationic heavy metals.

As detailed in section 6-3, such an electrostatic interaction has been proposed as the first step in intestinal metal absorption. Direct evidence for binding of Cd and Zn to membrane phospholipids in the intestinal brush border was reported by Tacnet et al. (1991) on the basis of nuclear magnetic resonance measurements. The density of anionic sites on cell membranes is quite high: Hanck and Sheets (1992), for instance, calculated a value of $0.72 \times 10^{14}/cm^2$ on canine cardiac Purkinje cells. Thethi and Duszyk (1997) found 1.2×10^{10} anionic charges per epithelioid cell from the human pancreas; this is reduced by about 40% following removal of sialic acid residues with neuraminidase.

In summary, a multiplicity of reactions occur at the membrane, illustrating the active role played by this structure in cell function. Not only do membrane constituents participate in these reactions, but the membrane itself is a dynamic rather than static structure that continuously turns over under suitable conditions. This so-called membrane traffic is discussed in the next section.

4-3 MEMBRANE FLUIDITY AND TURNOVER

Membranes continuously break up and reform as part of such normal physiologic processes as cell growth, endocytosis, exocytosis, cell division, and cell locomotion. Changes in cell volume are obviously associated with changes in membrane geometry and stretch (see Chapter 8). In addition, composition and function of membranes adapt to challenges such as changes in diet or the environment (see Chapter 7) and respond to certain drugs or selected physiologic messengers like hormones. In other words, membranes are dynamic rather that static structures.

Further, membranes are not homogeneous structures, as shown, for instance, by the presence of cholesterol domains (see review by Schroeder et al., 1996). Diffusion measurements on glycoproteins in membranes also suggest the existence of a mosaic structure (Sheetz, 1993). Singer and Nicolson (1972) had proposed a fluid mosaic model for the distribution of proteins in membrane lipid bilayers. The dynamic nature of cell membranes also is supported by the apparent clustering of sphingolipids and cholesterol to form "rafts" floating on the fluid bilayer (Simons and Ikonen, 1997).

Membranes may respond to changes in the cellular environment by fragmentation and reconstitution, as well as by altering their composition and consequently their fluidity (Stubbs and Smith, 1984). The importance of membrane fluidity in determining the permeability of membranes to xenobiotics is discussed in this chapter and in section 6-3. For instance, dietary substitution of unsaturated for saturated fatty acids leads to corresponding changes in the composition of enteric cell membranes, which consequently become more fluid in nature. A number of xenobiotics, as seen for instance by the effects of n-butanol in Fig. 2-3, also influence the fluidity of biologic membranes, and thereby their permeability properties. At a concentration of 1.5% (v/v) this alcohol greatly accelerates the rate of cadmium uptake by membrane vesicles from the rat jejunum, and at the same time makes all the accumulated metal accessible to the normally extravesicular chelator EDTA (Bevan and Foulkes, 1989).

The significance of membrane fluidity to membrane function was reviewed by Ochsner (1997), who noted that "the physical state, or microviscosity, of the plasma membrane has been implicated in altering the conformation of transmembrane ion channels and/or in changing the activity of enzymes in the phospholipid matrix." Any toxicant affecting fluidity is likely thereby to influence membrane permeability, as illustrated in Fig. 2-3 for cadmium. Specifically for heavy metals in the intestinal mucosa, and probably also in some other cell types, fluidity itself appears to be a controlling variable in transmembrane transport, as suggested by the temperature-dependence of passive uptake (see section 6-3). The correlation between metal permeability and membrane fluidity is seen also in the observations that the relatively high intrinsic metal permeability of the immature gut decreases with maturation (Foulkes, Mort, and Buncher, 1991), as does the fluidity of the intestinal brush border membrane (Israel et al., 1987).

The importance of fluidity in the definition of membrane permeability also was emphasized by Aungst (1996), who observed that compounds that enhance oral mucosal permeation of drugs raise the fluidity of epithelial lipids. Engelke, Tahti, and Vaalavirta (1997) along similar lines described the perturbation of artificial and biologic membranes by aliphatic, alicyclic, and aromatic compounds. The effect was monitored in terms of increased membrane fluidity: the higher the concentration of the agents, the higher the degree of structural disorder observed in the membrane. Micromorphologic changes in the membrane associated with exocytosis could be visualized with atomic force microscopy in living renal epithelial and other cells (Schneider et al., 1997).

Membrane traffic, consisting of complex physiologic processes of breakdown, rearrangement, and reconstitution of cell membranes, also may be sensitive to toxic action. The toxicologic significance of membrane traffic is emphasized in section 2-6. Hormones determining the transport of specific solutes across cell membranes, such as insulin or aldosterone, may act by stimulating synthesis of carrier proteins or their incorporation into the cell membrane (see, e.g., Beron and Verry, 1994). In the mammalian collecting duct and the amphibian bladder, antidiuretic hormone has been shown to induce transfer of water channels (aquaporins) from cytoplasmic vesicles to the apical cell membrane (Hays et al., 1994). Membrane trafficking in adipocytes from rat brown fat tissue is stimulated by the purinergic action of traces of ATP (Pappone and Lee, 1996); such dependence on ATP implies a potential sensitivity of membrane processes dependent on membrane fluidity to inhibition by metabolic poisons.

The magnitude of membrane traffic involved in the function of some cells can be very significant. This is the case, for instance, for the continuous endocytic renal reabsorption of filtered plasma albumin in the rat kidney (Cui et al., 1996). Like other membrane transport processes, albumin reabsorption must as a first step involve an interaction between the transported solute and some specific membrane constituent or constituents, in this case a glycoprotein (medalin). Extensive and rapid recycling of portions of cell membranes are indicated also in reports of intracellular accumulation of monoclonal antibodies following their binding to cell membranes. A further illustration of the dynamic nature of membrane function is provided by the results of Schwab et al. (1994), who observed that transformed Madin-Darby canine kidney (MDCK) cells in culture exhibit spontaneous oscillatory activity of potassium channels associated with cell locomotion.

A number of other instances also have been described frequently in which membrane-bound compounds are directly internalized. Thus, Bridges et al. (1982) found that, within 5 minutes of exposing isolated hepatocytes to asialoglycoprotein, the greater portion of this protein that had become bound to the cell membrane is transferred into the cells. Section 6-3 describes evidence for the internalization of heavy metals bound to the brush border membrane of intestinal epithelium.

The response of renal cells to hypertonic saline, or to mitogenic stimulation, includes an increased activity of the membrane Na^+/H^+ antiport (Schuldiner and Rozengurt, 1982). The activity of the system A neutral amino acid carrier in mammalian cell membranes (Chen and Kempson, 1995) also is stimulated by hyperosmotic media. These adaptive or compensatory responses may involve redistribution of existing transport proteins from intracellular sites to the outer cell membrane or induction of carrier synthesis, in which case the response may be sensitive to inhibitors of protein synthesis such as oligomycin. The rapidity with which relocation of such membrane constituents may be triggered by physiologic stimuli is illustrated in Fig. 4-1, which shows results of centrifugal density gradient fractionation of Na,K-ATPase in the proximal tubule of control and acutely hypertensive rats (Zhang et al., 1996).

Note that in control animals, peak activity (40% of total) was recovered in lower-density fractions (3,4) corresponding to basolateral membranes. Within 5 minutes of acutely initiating natriuresis in rats by raising arterial blood pressure, the ATPase activity was shifted toward higher-density fractions rich in markers of intracellular

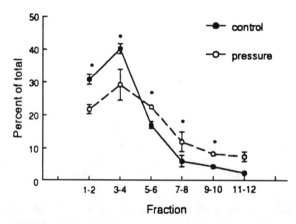

FIG. 4-1. Relocation of basolateral Na,K-ATPase in acutely hypertensive rats. (From Zhang et al., 1996; with permission.)

membranes such as acid phosphatase and β-hexosaminidase. A similar redistribution to higher-density membranes was observed for the brush border Na/H exchanger. Acute proximal tubular natriuresis, therefore, could be attributed, at least in part, to rapid endocytic removal of sodium transporters from the outer to intracellular membranes. Acting to stimulate membrane transport rather than to decrease it, the hormone insulin recruits intracellular glucose transporters for incorporation into the cell membrane; this explains both the insulin stimulation of cellular glucose uptake and the short but significant delay preceding this stimulation.

The discussion to this point has emphasized the extent of membrane traffic in normal cells. The sensitivity of such traffic to toxic interference is illustrated by the results of Blumenthal et al. (1990) on the cadmium inhibition of glucose uptake in primary cultures of mouse tubule cells. This effect presumably results from reduced intracellular synthesis or recruitment of glucose transporters; after 24 hours of exposure to micromolar concentrations of Cd, Na-glucose cotransport was depressed without overt effects on cell viability or energy metabolism. Measurement of phlorizin binding indicated that the action of Cd did reduce the number of functional glucose transporters on the membrane.

4-4 MEMBRANE PROTEINS

The lipoprotein matrix of the membrane is associated with many specific proteins responsible for membrane function. As alrady discussed, they include a variety of enzymes anchored in the matrix but with catalytic domains in the aqueous phase. If these enzymes catalyze extracellular reactions they are referred to as ectoenzymes; other membrane-bound proteins, such as the mammalian adenyl cyclases reviewed by Taussig and Gilman (1995), mediate reactions in the cells. Membrane proteins also include transmembrane molecules acting as channels and transporters. In addition, many complex membrane functions such as cell recognition, communication, and adhesion are carried out by complex protein molecules generally containing sugars, sialic acid, and other residues. A widely studied group of proteoglycans, the integrins, are the primary membrane receptors for interacting with the extracellular matrix (for review, see Burridge and Chrcwanowska-Wodnicka, 1996). Another group of membrane-associated proteins with special functions are the annexins (Kaetzel and Dedman, 1995), involved in Ca^{2+}-dependent cell regulation by modifications of membrane function.

The toxicologic relevance of such membrane-associated proteins can be illustrated, as mentioned in section 4-1, by the effects of lead on the function of the neurocellular adhesion molecules believed to be intimately involved in the orderly structuring of the developing central nervous system (Cookman, King, and Regan, 1987). The rationale underlying these experiments was the suggestion that the extent of sialylation (formation of sialic acid complexes) of adhesion molecules helps regulate cell–cell interaction, fiber outgrowth, and synapse formation. Rat pups in these studies therefore were exposed to Pb in drinking water through their mothers' milk. At age 16

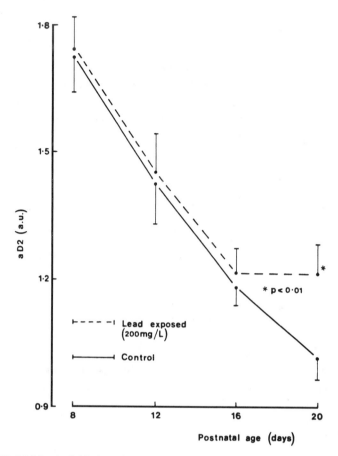

FIG. 4-2. Pb inhibits desialylation of neural adhesion molecules. Concentration of the sialic acid–rich fraction aD2 is shown in arbitrary units during postnatal development. (Adapted from Cookman, King, and Regan, 1987; with permission.)

days, the blood lead in the exposed pups began to exceed 18 μg/dL; at the same time point, as shown in arbitrary units in Fig. 4-2, the normal desialylation in control animals came to a stop in the lead-treated animals, so that the concentration of the sialic acid–rich fraction (aD2) decreased no further. This inhibition may be associated with the neurobehavioral deficits observed in developing animals (and humans) exposed to lead.

The toxicologic significance of membrane-associated proteins is seen further in the many instances in which toxic agents cause their release from the membrane. Thus, damage to the proximal brush border in the kidney increases the urinary excretion of such specific membrane proteins as adenosine deaminase–binding protein, Na^+-glucose cotransporter protein, intervillus proteins, and intestinal-type alkaline phosphatase (Birk et al., 1991). The urinary excretion of N-acetylglucosaminidase and other enzymes frequently is used to monitor renal damage.

Another membrane-associated protein whose urinary excretion may be of diagnostic importance is the so-called Tamm-Horsfall glycoprotein. Its release from apical membranes in the renal tubule by proteolytic or phospholipolytic cleavage (Rindler et al., 1990) indicates that it is one of the proteins bound to the membrane by direct reaction with phospholipids, rather than on the basis of hydrophobic peptide regions sequestered in the lipid layers.

The nature of protein binding to membranes became much clearer with the elucidation of primary protein structure. This permitted the identification in many proteins of hydrophobic domains in the peptide chain, likely to be located in, and thus anchoring the protein to, the lipid membrane. It follows that the catalytic or specific binding sites for water-soluble and extracellular substrates and agonists are to be found in the hydrophilic sections of receptors and enzymes on the membrane, stretching out from the matrix into the aqueous phase. Photosensitive hydrophobic reagents labeled with ^{125}I have proven particularly useful for labeling the anchoring portion of membrane proteins (Booth, Hubbard, and Kenny, 1979). A single protein with multiple hydrophobic domains thus may loop into and out of the cells a number of times, with the amino and carboxy ends of the peptide chain freely accessible in the aqueous phase. Such structures have been established for many membrane-associated proteins. Other proteins, which do not contain significant hydrophobic domains, may be bound directly to membrane phospholipids and similar structures, as indicated previously (Luzio, Baron, and Bailyes, 1987).

In addition to their anchoring function, transmembrane domains of proteins also may play a role in transmembrane signal transduction. This is true, for instance, of the agonist receptors that transmit signals to second messengers inside the cells. Signal transmission across membranes commonly depends on the presence of guanosine triphosphate-binding G proteins bound to the cell membranes; another group of proteins involved in transmembrane processes is electron transfering enzymes (Kilberg and Christensen, 1979). The transmembrane redox system in Ehrlich ascites tumor cells was identified as a glycoprotein whose function is influenced by changes in its phospholipid environment (Del Castillo-Olivares et al., 1994). Such redox enzymes are responsible for many functions, including solute transport and signal transduction. Crane et al. (1985) suggested that they also might possess a role in such physiologic processes as growth and development.

Enzymes that are anchored in the cell membranes but catalyze reactions in the extracellular phase are referred to as *ectoenzymes* (Kreutzberg, Reddington, and Zimmerman, 1985), a term originally proposed by Engelhardt (1957) to describe the catalyst responsible for the rapid hydrolysis of ATP by avian erythrocytes. They are integral membrane proteins, commonly solubilized by detergent treatment, limited proteolysis, or the type of toxic membrane damage causing excretion of tubular brush border enzymes in urine (see subsequent text). Numerous ectoenzymes have been described, specific for a wide variety of substrates; some representative examples from mammalian cells (based on Kenny and Turner, 1987) include:

gamma-glutamyl transferase
nucleoside diphosphate kinase

acetylcholineesterase
alkaline phosphatase
5′-nucleotidase
glucoamylase–maltase
lactase–glycosylceramidase
trehalase
sucrase–isomaltase
aminopeptidases
carboxypeptidase
pyroglutamyl peptidase II
enteropeptidase
endopeptidases
neurotensin-degrading enzyme
nucleoside diphosphatase
nucleoside triphosphatase

The potential toxicologic significance of these enzymes is indicated by their sensitivity to various xenobiotics. Kenny, Stephenson, and Turner (1987), for instance, listed a whole series of specific inhibitors of microvillar membrane peptidases; general protein reagents such as Hg^{2+}, of course, also will depress activity.

Numerous other specific proteins and enzymes have been found associated with cell membranes. They include, especially, the transport systems mediating solute movement across the membrane (see sections 4-6 and 4-7 below). Although some of them are sensitive to highly specific inhibitors, their activity generally is abolished by heavy metals and other nonspecific protein reagents (see Chapter 5). A number of the transport systems act in association with hydrolysis of ATP. Examples of ATPases thus involved in ion transport include the Na,K-dependent ATPase generally mediating Na extrusion from cells, a Ca,Na-ATPase participating in Ca transport, a Na,H-ATPase involved in gastric acid secretion, and a chloride-requiring ATPase catalyzing cellular chloride transport.

Several membrane transporters have been purified and their action studied in reconstituted lipid membranes. Examples of such studies are the extensive work on glucose transporters (James, 1995) and the characterization of Na,K-ATPase. Similarly, Na^+-independent transport of organic anions in hepatocytes is associated with a discrete glycoprotein of molecular mass 74 kD and 670 amino acid residues (Hagenbuch, Jaquemin, and Meier, 1994). The P-glycoprotein mediating the export of toxic xenobiotics from cells (see section 7-3) also has been well characterized.

In addition to specific transport systems mediating the facilitated or active transfer of selected solutes across the membrane, a variety of membrane channels, as defined and summarized in sections 4-6 and 4-7, facilitate transmembrane movement of small molecules and ions, including water, urea, and electrolytes. Opening and closing of these channels often are controlled by intricate gating mechanisms, involving such variables as concentration of ATP or other specific solutes, and transmembrane voltages; inhibitors specifically depressing channel function have been described. An example of such a channel is the so-called protein band III from erythrocyte membranes.

This is a glycoprotein consisting of 911 amino acid residues and 14 transmembrane domains (Lux et al., 1989). It has been investigated extensively in many laboratories and is responsible for the high anion permeability of these cells. The classic anion channel blocker diisothionatostilbene disulfonate (DIDS) reacts with band III.

Given the dynamic structure of membranes (see section 4-3), together with their association with all these proteins that either act as structural components or are involved dynamically in membrane function, it is not surprising to find that toxicants possessing high general affinity for proteins readily and nonspecifically interfere with membrane function. Mercuric mercury, for instance, depresses activity of several easily accessible membrane enzymes (see sections 2-4 and 4-5). Acute in-vitro effects of mercury on the membrane range from inhibition of the water channel aquaporin (Preston et al., 1993) to depression of Na,K-ATPase (Nechay and Saunders, 1978) and interference with glucose uptake (see Chapter 2). Multiple effects of the metal on transport of L-alanine by membrane vesicles are illustrated in Table 5.1. It is difficult in many instances to decide whether these in-vitro observations can be directly extrapolated back to the intact organism and thus help define the primary targets of the metal in vivo (see Chapter 3).

More specific inhibitors of membrane enzymes also have been identified (Chapter 5). The classic example is provided by the action of ouabain on Na,K-ATPase. Specific inhibitors of various ectoenzymes (ectopeptidases) are listed in Kenny and Turner (1987). It must be remembered that primary inhibition of a membrane enzyme may lead to general cytotoxicity if it interferes, for example, with cell homeostasis.

Nature and function of membrane proteins constituting receptors, transporters, and channels are further summarized here under separate headings.

4-5 RECEPTORS

The ability of cells to respond to external stimuli often is associated with specific receptors on the cell membrane. They are protein in nature and react directly with a variety of extracellular agonists and stimuli including many drugs, hormones, and neurotransmitters ("first messengers"). Other first messengers bind to intracellular receptors such as the cytoplasmic aryl hydrocarbon receptor. First messengers reaching this toxicologically very important receptor are generally lipid-soluble and accordingly are able passively to penetrate the cell.

Numerous receptor agonists and antagonists have been identified, able to simulate the reaction of receptors with first messengers or to block the receptors. Such reactions recently have attracted special attention because of the apparent estrogen-like activity of many pollutants. Here is not the place, however, for a detailed discussion of the rapidly expanding field of receptor pharmacology or of the mechanism whereby messages are transferred (message transduction) from receptors to effector sites. The topic is cited here because a number of toxic xenobiotics have been identified as receptor blockers whose reaction with membrane receptors further illustrates the complexity of membrane function.

Membranes also respond to mechanical stimuli like stretch. Hoyer et al. (1996), for instance, described a mechanosensitive Ca channel in the vascular epithelium of hypertensive rats, which is activated on application of a positive pressure through patch-clamp pipettes. Stretch-sensitive membrane solute channels contribute to the prevention of cell swelling in hypotonic solutions, as further discussed in Chapter 8. The transmembrane proteins acting as osmoreceptors in yeast (Maeda, Takekawa, and Saito, 1995) may react to changes in turgor pressure rather than volume, as swelling of the yeast cell is prevented by a rigid cell wall. The precise nature of the response of membrane to mechanical factors remains unclear, although a variety of mechanisms have been proposed for mechanochemical signal transduction.

Binding of an agonist to a membrane receptor may lead to allosteric changes in the receptor (Chervitz and Falke, 1996), or a message may be transferred across the membrane by some other mechanism. Generally, such transfer triggers a cascade of further reactions leading to changes in the cytosolic concentration or activity of "second messengers" like cyclic AMP or inositol triphosphate. Stimulation of such a second-messenger response may be evoked at the cell membrane itself by, for instance, phospholipase-catalyzed breakdown of membrane phosphatide and production of inositol triphosphate.

Membrane receptors have been identified and cloned for a number of hormones including growth hormone, prolactin, erythropoietin, and various cytokines (Hochberg, 1994). Specific membrane-binding proteins (receptors) have been described for thyroid hormone (Whelan and Hill, 1993), dexamethasone (Howell, Po, and Lefebvre, 1989), methotrexate (Antony et al., 1985) and a number of other agonists. Of particular toxicologic importance are acetylcholine receptors at nerve synapses (see section 5-3). Histamine receptors also have been studied in considerable detail. Gantz et al. (1991) reported the molecular cloning of a gene encoding the histamine-2 receptor.

4-6 SOLUTE AND WATER CHANNELS AND PORES

Another class of specific membrane proteins consists of the solute channels that determine the permeability of the membrane to many lipid unsoluble substances. These channels generally may be distinguished from simple aqueous pores by the facts that they are specific for certain solutes and under homeostatic control. Channels have been identified for various inorganic and organic solutes. The generally accepted model of these channels visualizes an aqueous pore whose walls are formed by the proteins; these, in turn, determine the specificity of the channels. Unlike the specific ionophores, for example, for Ca^{++}, Na^+, and K^+, which are smaller organic molecules whose ability to increase ion conductance across the membrane is not physiologically regulated, the function of relatively specific channels is subject to feedback control. For instance, ion channels in the cell membrane can serve as volume or stretch receptors, thus permitting cells to alter their solute permeability in response to a physiologic stimulus such as cell swelling (see section 8-1).

A further example of sensitively controlled channels is that of the various Ca channels in excitable cells, which are voltage-gated and activated by membrane depolarization. In certain cell types, such as pituitary secretory cells (Hinkle, Kinsella, and Osterhoudt, 1987), such Ca channels may mediate uptake of Cd (see section 6-3). In general, channels may be classified on the basis of their substrate specificity and of their sensitivity to particular organic and inorganic channel blockers. They also differ, as already mentioned, in their gating behavior, or the factors and characteristics determining when and for how long they remain in the open or closed state. Thus, Sperelakis (1995) listed the following gating mechanisms of ion channels in membranes of cardiac myocytes: voltage gating, ligand gating as for the ion channels at the neuromuscular junction blocked by acetyl choline, G-protein gating, Ca^{++} or Na^+ activation, ATP concentration, cytoplasmic levels of cyclic nucleotides, and stretch sensitivity.

Activity of solute channels may be inhibited by a variety of toxic agents or channel blockers. Barium ions, for instance, block certain K channels. Natural toxins blocking Na channels are listed in Table 5-3. Mercurials, as reviewed by Sirois and Atchison (1996), inhibit ligand- and voltage-gated ion channels in neuronal cell membranes. As nerve cells rely on ion channels in the membrane for many of their primary functions, the cell membrane represents an obvious target for neurotoxicants.

Presence of aqueous or polar pores in membranes has long been recognized and widely studied. No attempt is made here to review the older work that deduced equivalent pore diameters in membranes from permeability characteristics to water-soluble compounds. The ability of urea to cross cell membranes has been used as indicator of the frequency of aqueous pores (Macey, 1984). In line with the dynamic nature of membranes (see sections 2-6 and 5-3), the concentration or frequency of membrane pores is not constant but responds to a variety of external factors. An example of such a response, in this case presumably mechanical stress, is the observation that uptake of urea by jejunal epithelium in everted sacs of intestine exceeds that in the intact tissue (Foulkes and Bergman, 1993). As indicated in section 3-5, this represents a good example of an artifact produced during the preparation of an in-vitro model.

In addition to these polar pores in the membranes, intercellular diffusion pathways also can contribute to solute movement across the intestinal barrier. Such pathways, for instance, account for the movement of phenol red across the mucosa of the small intestine (Tidball, 1964). The movement is stimulated by EDTA, suggesting a role of Ca in limiting the permeability of tight junctions between cells. Actions of certain toxins also affect the general permeability of the intestinal barrier. Thus, an enterotoxin from *Vibrio cholerae* (Fasano et al., 1991) increases intestinal tissue conductance by altering the structure of intercellular tight junctions; a toxin from *Clostridium difficile* (Moore et al., 1990) also reduces the barrier function of intestinal epithelium. The mechanism of this action is not clearly understood but may involve effects on the cytoskeleton.

Besides such mechanical or drug-induced changes in the general permeability of cells or cell barriers, specific hormone effects on passive cell permeability frequently have been described. The classic instance is the well-known physiologic variations in the water permeability of membranes in the distal nephron under the influence of the antidiuretic hormone. Although biologic membranes generally exhibit at least a small permeability to H_2O, portions of the renal tubule, the amphibian urinary bladder, and the red blood cell are distinguished by extremely high water fluxes. Transmembrane water-channel proteins are involved in water movement across many tissues (Bachinsky et al., 1995). Flow of water reflects presence of special water pores called *aquaporins* (Verkman, 1989); these are specific proteins of which one (CHIP28) has been expressed in *Xenopus laevis* oocytes (Preston et al., 1992); like so many membrane proteins, aquaporin is sensitive to mercury.

Human aquaporin channels were reported by Abrami, Tacnet, and Ripoche (1995) to facilitate transmembrane movement of low–molecular weight polyalcohols like ethylene glycol, glycerol, and 1,3-propane diol. The density of aquaporins in the membrane was estimated at 2×10^5 per red cell. By way of comparison, Hoffmann, and Simonsen (1989) estimated on the indirect basis of membrane conductance that the density of Ca-activated K^+ channels on Ehrlich ascites cells is about 100 per cell. The density of sodium channels in frog-skin epithelium has been reported as 400 per square micrometer (Cuthbert, 1973).

4-7 SOLUTE TRANSPORTERS

The distinction between channels and transporters (or pumps) is not always defined clearly. For instance, the P-glycoprotein discussed in Chapter 7 exhibits properties of a chloride and an ATP channel, as well as those of a solute pump. In general, transacceleration and countertransport typically reflect carrier activity, as does a slower turnover rate than that of channels (Ballatori, Simmons, and Boyer, 1994). Carriers may be distinguished further from pores on the basis of kinetic evidence, as described by Yousef and Macey (1989) for urea transport.

As already emphasized in section 4-4, a great variety of solute transporters have been identified on cell membranes and are indeed essential for the supply of metabolic substrates and the extrusion of waste or other unneeded or toxic materials, and thereby for the maintenance of steady-state cell composition. The direct or indirect effects on these transport proteins exerted by various xenobiotics is discussed in Chapter 5.

Such actions may represent the desired outcome of drug therapy or the effect of a toxic agent. Obviously the identification of specific therapeutic or toxic agents directly reacting with transporters is likely to provide much more information on the nature and function of solute transport, and the mechanism of action of toxicants in vivo, than can the effects of general inhibitors like mercury. Specific inhibitors may act competitively, as in the case of the interaction between aspartate and glutamate for amino acid transporters, or noncompetitively, as in the case of the ouabain inhibition of Na,K-ATPase. Chapter 5 provides a more detailed discussion of the nature of toxic interaction between xenobiotics and membrane function.

4-8 CONCLUSION

This chapter summarizes salient aspects of membrane composition and function; its objective is to emphasize the complexity of both these parameters. A comprehensive discussion of membrane science lies beyond the scope of this work.

The general accessibility of membranes to extracellular xenobiotics, and the complex structure required for membranes to be able to carry out their functions, predictably make them a frequent target of toxic action. Indeed, a number of cytotoxic effects have been correlated with more or less specific functional lesions at the cell membrane. As further discussed in Chapter 5, such lesions may result directly from actions at the membrane; alternatively, they may arise indirectly from reactions elsewhere in the cell or in the body.

Membrane processes are involved closely in the absorption especially of nonlipophilic toxic agents into the body, their distribution between various body compartments, their access to intracellular targets and their metabolism, their cellular extrusion, and finally their primarily renal or hepatic excretion. These facts, together with the basics of membrane science discussed in this chapter, provide explanations for two important aspects of membrane toxicology: the sensitivity of membranes to the direct or indirect action of many toxic xenobiotics (see Chapter 5), and the determining role of cell membranes in the toxicokinetics of nonlipophilic xenobiotics (Chapter 6-2).

REFERENCES

Abrami, L., F. Tacnet, P. Ripoche. 1995. Evidence for a glycerol pathway through aquaporin I (CHIP28) channels. *Pflugers Arch.* 430:447–458.

Antony, A.C., M.A. Kane, R.M. Portillo, P.C. Elwood, and J. F. Kolhouse. 1985. Studies on the role of a particulate folate-binding protein in the uptake of 5-methyltetrahydrofolate by cultured human KB cells. *J. Biol. Chem.* 260:14911–14917.

Aungst, B. 1996. Oral mucosal permeation enhancement. In *Oral mucosal drug delivery*, ed. M.J. Rathbone, 65–83. New York: Marcel Dekker.

Bachinsky, D.R., I. Sabolic, D.S. Emmanouel, D.M. Jefferson, F.A. Carone, D. Brown, and R.D. Perrone. 1995. Water channel expression in human ADPKD kidneys. *Am. J. Physiol.* 268:F398–F403.

Ballatori, N., T.W. Simmmons, and J.L. Boyer. 1994. A volume-activated taurine channel in skate hepatocytes: Membrane polarity and role of intracellular ATP. *Am. J. Physiol.* 267:G285–G291.

Beron, J., and F. Verry. 1994. Aldosterone induces early activation and late accumulation of Na,K-ATPase at surface of A6 cells. *Am. J. Physiol.* 266:C1278–C1290.

Bevan, C., and E.C. Foulkes. 1989. Interaction of cadmium with brush border membrane vesicles from the rat small intestine. *Toxicology* 54:297–309.

Birk, H.W., S. Piberhofer, G. Schutterle, W. Haase, J. Kotting, and H. Koepsell. 1991. Analysis of Na,D-glucose cotransporter and other renal brush border proteins in human urine. *Kidney Int.* 40:823–837.

Blumenthal, S.S., D.L. Lewand, M.A. Buday, J.G. Kleinman, S.K. Krezoski, and D.H. Petering, 1990. Cadmium inhibits glucose uptake in primary cultures of mouse cortical tubule cells. *Am. J. Physiol.* 258:F1625–F1633.

Bodenstab, W., J. Kaufman, and C.L. Parsons. 1983. Inactivation of anti-adherence effect of bladder surface glycosaminoglycan by a complete urinary carcinogen (N-methyl-N-nitrosourea). *J. Urol.* 129:200–201.

Booth, A.G., L.M. Hubbard, and A.J. Kenny. 1979. Proteins of the kidney microvillar membrane. *Biochem. J.* 179:397–405.

Bridges, K., J. Harford, G. Ashwell, and R.D. Klausner, 1982. Fate of receptor and ligand during endocytosis of asialoglycoproteins by isolated hepatocytes. *Proc. Natl. Acad. Sci. U.S.A.* 79:350–354.

Burridge, K., and M. Chrzanowska-Wodnicka. 1996. Focal adhesions, contractility and signaling. *Ann. Rev. Cell Dev. Biol.* 12:463–518.

Byrne, J.H., and S.G. Schultz. 1997. *An introduction to membrane transport and bioelectricity.* Hagerstown, MD: Lippincott-Raven.

Chen, J.G., and S.A. Kempson. 1995. Osmoregulation of neutral amino acid transport. *Proc. Soc. Exp. Biol. Med.* 210:1–6.

Chervitz, S.A., and J.J. Falke. 1996. Molecular mechanism of transmembrane signaling by the aspartate receptor: A model. *Porc. Natl. Acad. Sci. U.S.A.* 93:2545–2550.

Cookman, G.R., W. King, and C.M. Regan. 1987. Chronic low-level lead exposure impairs embryonic to adult conversion of the neural cell adhesion molecule. *J. Neurochem.* 49:399–403.

Crane, F.L., I.L. Sun, M.G. Clark, C. Grebing, and H. Low. 1985. Transplasma-membrane redox systems in growth and development. *Biochim. Biophys. Acta.* 811:233–264.

Cui, S., P.J. Verroust, S.K. Moestrup, and E.I. Christensen. 1996. Megalin/gp330 mediates uptake of albumin in renal proximal tubule. *Am. J. Physiol.* 271:F900–F907.

Cuthbert, A.W. 1973. An upper limit to the number of sodium channels in frog skin epithelium. *J. Physiol.* 228:681–692.

Davson, H., and J.F. Danielli. 1952. *The permeability of natural membranes.* Cambridge, UK: Cambridge University Press.

Del Castillo-Olivares, A., J. Marquez, I.N. De Castro, and M.A. Medina. 1994. Characterization of plasma membrane redox activity from Ehrlich cells. *Cell Biochem. Function* 12:149–152.

Engelhardt, W.A. 1957. Enzymes as structural elements of physiological mechanism. *Proc. Int. Symp. Enzyme Chem. (Tokyo)* 2:163–166.

Engelke, M., H. Tahti, and L. Vaalavirta. 1997. Perturbation of artificial and biological membranes by organic compounds of aliphatic, alicyclic and aromatic structure. *Toxicol. in Vitro* 10:111–115.

Fasano, A., B. Baudry, D.W. Pumplin, S.S. Wasserman, B.D. Tall, J.M. Ketley, and J.B. Kaper. 1991. V. cholerae produces a second enterotoxin which affects intestinal tight junctions. *Proc. Natl. Acad. Sci U.S.A.* 88:5242–5246.

Foulkes, E.C., and D. Bergman. 1993. Inorganic Hg absorption in mature and immature rat jejunum: Transcellular and intercellular pathways in vivo and in everted sacs. *Toxicol. Appl. Pharmacol.* 120:89–95.

Foulkes, E.C., T. Mort, and R. Buncher. 1991. Intestinal cadmium permeability in mature and immature rats. *Proc. Soc. Exp. Biol. Med.* 197:477–481.

Gantz, I., M. Schaffer, J. DelValle, C. Logsdon, V. Campbell, M. Uhler, and T. Yamada, 1991. Molecular cloning of a gene encoding the histamine H_2 receptor. *Proc. Natl. Acad. Sci. U.S.A.* 88:429–433.

Hagenbuch, B., E. Jaquemin, and P.J. Meier. 1994. Na-dependent and Na-independent bile acid uptake system in the liver. *Cell Physiol. Biochem.* 4:198–205.

Hanck, D.A., and M.F. Sheets. 1992. Extracellular divalent and trivalent cation effects on sodium current kinetics in single canine cardiac Purkinje cells. *J. Physiol.* 454:267–298.

Hays, R.M., N. Franki, H. Simon, and Y. Gao. 1994. Antidiuretic hormone and exocytosis: Lessons from neurosecretion. *Am. J. Physiol.* 267:C1507–1524.

Hinkle, P.M., P.A. Kinsella, and K.C. Osterhoudt. 1987. Cadmium uptake and toxicity via voltage-sensitive calcium channels. *J. Biol. Chem.* 262:16333–16337.

Hochberg, Z. 1994. The super family of receptors for growth hormone, prolactin, erythropoietin and cytokines: Introduction. *Proc. Soc. Exp. Biol. Med.* 206:169.

Hoffmann, E.K., and L.O. Simonsen. 1989. Membrane mechanisms in volume and pH regulation in vertebrate cells. *Physiol. Rev.* 69:315–382.

Howell, G.M., C. Po, and Y.A. Lefebvre. 1989. Identification of dexamethasone-binding sites on male-rat liver plasma membranes by affinity labelling. *Biochem. J.* 260:435–441.

Hoyer, J., R. Kohler, W. Haase, A. Distler. 1996. Upregulation of pressure-activated Ca^{2+}-permeable cation channel in intact vascular endothelium of hypertensive rats. *Proc. Natl. Acad. Sci. U.S.A.* 93:11253–11258.

Israel, E.J., K.Y. Pang, P.R. Harmutz, W.A. Walker. 1987. Structural and functional maturation of rat gastrointestinal barrier with thyroxine. *Am. J. Physiol.* 252:G762–G767.

James, D.E. 1995. The mammalian facilitative glucose transporter family. *News Physiol. Sci.* 10:67–71.

Kaetzel, M.A., and J.R. Dedman. 1995. Annexins: Novel Ca^{2+}-dependent phospholipid binding proteins that alter membrane properties and cellular function. *News Physiol. Sci.* 10:171–176.

Kenny, A.J., S.L. Stephenson, A.J. Turner. 1987. Cell surface peptidases. In *Mammalian ectoenzymes*, eds. A.J. Kenny and A.J. Turner, 169–210. Amsterdam: Elsevier.

Kenny, A.J., and A.J. Turner. 1987. What are ectoenzymes? In *Mammalian ectoenzymes*, eds. A.J. Kenny, and A.J. Turner, 1–13. Amsterdam: Elsevier.

Khatri, I.A., G.G. Forstner, J.F. Forstner. 1997. The carboxy-terminal sequence of rat intestinal mucin RMuc3 contains a putative transmembrane region and two EGF-like motifs. *Biochim. Biophys. Acta* 1326:7–11.

Kilberg, M.S., and H.N. Christensen. 1979. Electron transferring enzyme in the plasma membrane of the Ehrlich ascites tumor cell. *Biochemistry* 18:1525–1530.

Kreutzberg, G.W., M. Reddington, and H. Zimmerman, 1985. *Cellular biology of ecto-enzymes*. Berlin: Springer Verlag.

Lux, S.E., K.M. John, R.R. Kopito, and H.F. Lodish. 1989. Cloning and characterization of band 3, the human erythrocyte anion-exchange protein. *Proc. Natl. Acad. Sci. U.S.A.* 86:9089–9093.

Luzio, J.P., M.D. Baron, and E.M. Bailyes. 1987. Cell biology. In *Mammalian Ectoenzymes*, eds. A.J. Kenny and A.J. Turner, 111–138. Amsterdam: Elsevier.

Macey, R.I. 1984. Transport of water and urea in red blood cells. *Am. J. Physiol.* 246:C195–C203.

Maeda, T., M. Takekawa, and H. Saito, 1995. Activation of yeast PBS2 MAPKK by MAPKKKs or by binding of an SH3-containing osmosensor. *Science* 269:554–558.

Moore, R., C. Pothoulakis, J.T. LaMont, S. Carlson, and J.L. Madara. 1990. C. difficile toxin A increases intestinal permeability and induces chloride secretion. *Am. J. Physiol.* 259:G165–G172.

Nechay, B.R., and J.P. Saunders. 1978. Inhibitory characteristics of cadmium, lead and mercury in human sodium and potassium dependent adenosinetriphosphatase preparations. *J. Environ. Pathol. Toxicol.* 2:283–290.

Nimmerfall, F., and J. Rosenthaler. 1980. Significance of the goblet cell mucin layer, the outermost luminal barrier to passage through the gut wall. *Biochim. Biophys. Acta* 94:960–966.

Ochsner, M. 1997. Ca^{2+} transient, cell volume, and microviscosity of the plasma membrane in smooth muscle. *Biochem. Pharmacol.* 53:1765–1777.

Pappone, P.A., and S.C. Lee. 1996. Purinergic receptor stimulation increases membrane trafficking in brown adipocytes. *J. Gen. Physiol.* 108:393–404.

Preston, G.M., T.P. Carrol, W.B. Guggino, and P. Agre. 1992. Appearance of water channels in xenopus oocytes expressing red cell CHIP-28 proteins. *Science* 256:385–387.

Preston, G.M., J.S. Jung, W.B. Guggino, P. Agre, 1993. The mercury-sensitive residue at cysteine 189 in the CHIP-28 water channel. *J. Biol. Chem.* 268:17–20.

Reuhl, K.R., and P.M. Dey. 1996. Cell adhesion molecules in metal neurotoxicity. In *Toxicology of Metals*, ed. L. Chang, 1097–1110. Boca Raton, FL: Lewis Publishers.

Rindler, M.J., S.S. Naik, N. Li, T.C. Hoops, and M.N. Peraldi. 1990. Uromodulin (Tamm-Horsfall glycoprotein/uromucoid) is a phosphatidylinositol-linked membrane protein. *J. Biol. Chem.* 265:20784–20789.

Schneider, S.W., K.C. Sritharan, J.P. Geibel, H. Oberleithner, and B.P. Jena. 1997. Surface dynamics in living acinar cells imaged by atomic force microscopy: Identification of plasma membrane structures involved in exocytosis. *Proc. Natl. Acad. Sci. U.S.A.* 94:316–321.

Schroeder, F., A.A. Frolov, E.J. Murphy, P.P. Atshaves, J.R. Jefferson, L. Pu, W.G. Wood, W.B. Foxworth, and A.B. Kier. 1996. Recent advances in membrane cholesterol domain dynamics and intracellular cholesterol trafficking. *Proc. Soc. Exp. Biol. Med.* 213:150–177.

Schuldiner, S., and E. Rozengurt. 1982. Na^+/H^+ antiport in Swiss 3T3 cells: Mitogenic stimulation leads to cytoplasmic alkalinization. *Proc. Natl. Acad. Sci. U.S.A.* 79:7778–7782.

Schwab, A., L. Woynowski, K. Gabriel, and H. Oberleithner. 1994. Oscillating activity of a Ca^{2+}-sensitive K^+ channel. *J. Clin. Invest* 93:1631–1636.

Sheetz, M.P. 1993. Glycoprotein motility and dynamic domains in fluid phase plasma membranes. *Ann. Rev. Biophys. Biomol. Struct.* 22:417–431.

Simons, K., and E. Ikonen. 1997. Functional rafts in cell membranes. *Nature* 387:569–572.

Singer, S.J., and G.L. Nicolson. 1972. The fluid mosaic model of the structure of cell membranes. *Science* 175:720–731.

Sirois, J.E., and W.D. Atchison. 1996. Effects of mercurials on ligand- and voltage-gated ion channels: A review. *Neurotoxicol.* 17:63–84.

Sperelakis, N. 1995. *Cell physiology*. San Diego: Academic Press.

Stubbs, C.D., and A.D. Smith. 1984. The modification of mammalian membrane polyunsaturated fatty acid composition in relation to membrane fluidity and function. *Biochim. Biopys. Acta* 779:89–137.

Tacnet, F., P. Ripoche, M. Roux, J.M. Neumann. 1991. Phosphorus-31 NMR studies of pig intestinal brush border membrane structure: Effect of zinc and cadmium ions. *Eur. Biophys. J.* 19:317–322.

Taussig, R., and Gilman, A.G. 1995. Mammalian membrane-bound adenylyl cyclases. *J. Biol. Chem.* 270:1–4.

Thethi, K., and M. Duszyk. 1997. Decreased cell surface charge in cystic fibrosis epithelia. *Cell. Biochem. Funct.* 15:35–38.

Tidball, C.S. 1964. Magnesium and calcium as regulators of intestinal permeability. *Am. J. Physiol.* 206:243–246.

Verkman, A.S. 1989. Mechanisms and regulation of water permeability in renal epithelia. *Am. J. Physiol.* 257:C837–C850.

Whelan, R.D.H., and B.T. Hill. 1993. Differential expression of steroid receptors, hsp27 and pS2, in a series of drug resistant human breast tumor cell lines derived following exposure to antitumor drugs or to fractionated X-irradiation. *Breast Cancer Res. Treatment.* 26:23–39.

Yousef, L.W., and R.I. Macey. 1989. A method of distinguishing between pore and carrier kinetics applied to urea transport across the erythrocyte membrane. *Biochim. Biophys. Acta* 984:281–288.

Zhang, Y., A.K. Mircheff, C.B. Hensley, C.E. Magyar, D.G. Warnock, R. Chambrey, K.P. Yip, D.J. Marsh, N.H. Holstein-Rathlou, and A.A. McDonough. 1996. Rapid redistribution and inhibition of renal sodium transporters during acute pressure natriuresis. *Am. J. Physiol.* 270:F1004–F1014.

5

Toxic Effects on Membrane Structure and Function

5-1 INTRODUCTION

The preceding chapter described the presence in membranes of many more-or-less specific and reactive proteins and other macromolecules; further, membranes were shown to play critical roles in many essential cell processes. It is not surprising, therefore, to find that cell membranes often provide the target sites for toxic xenobiotics. This chapter deals with the nature of structural and functional membrane lesions resulting directly or indirectly from actions of toxic agents.

The reactions of toxic agents with membranes can be quite complex. In considering the mechanisms underlying toxic actions, it is important to attempt to distinguish between lesions directly induced at a given site, in contrast to those resulting indirectly from reactions elsewhere in the cell or even in the body. This distinction, as is seen subsequently, cannot always be drawn readily. Further, some toxic effects specifically may alter a particular membrane function, as in the case of the ouabain inhibition of sodium transport. In other instances, agents nonspecifically may interfere with general membrane integrity. For instance, reaction of the membrane with a general protein reagent like Hg^{++} may be expected nonspecifically to inhibit the action of several membrane-associated enzymes. Similarly, peroxidation of membrane lipids, or hydrolysis of membrane phospholipids, can affect broadly a number of processes at the membrane.

Most of the evidence reviewed in the first six sections of this chapter describes functional lesions at outer cell membranes. A number of toxic lesions, however, also have been found at mitochondrial and other subcellular membranes (see section 5-7). Finally, the essential role of cell membranes in immune phenomena logically leads in section 5-8 to a summary of the membrane as the target site of immunotoxic actions.

5-2 NATURE OF TOXIC MEMBRANE LESIONS

As pointed out, the ability of a toxic agent to influence membrane functions may result from its direct reaction with membrane constituents. Alternatively, membrane malfunction may represent a consequence of primary reactions at sites removed from the membrane (see section 2-3); in these instances, functional changes at the membrane represent a secondary or indirect effect. It actually is frequently difficult to draw a clear distinction between suspected primary versus secondary, or direct versus indirect, interference with normal function. This is further considered in section 5-3.

Toxic lesions produced directly at the membrane may range from highly specific to very general. A very specific action, for instance, is caused by ouabain, the classic inhibitor of the membrane-associated Na,K-ATPase involved in sodium transport. This inhibition, however, leads to dissipation of the sodium gradient required for the active uptake of, for example, glucose. There is no evidence here for a direct role of the ATPase in glucose transport, and the depression of this transport by ouabain represents no more than an indirect effect. The ouabain inhibition of ascorbic acid uptake by human lymphoblasts (Ngkeekwong and Ng, 1997) can be explained in a similar manner. Both examples illustrate the difficulties that may be encountered in attempting to distinguish between direct and indirect toxic actions on membrane function. This is discussed further in section 5-3.

Alternatively, the effect of a toxic agent may consist of a more general attack on membrane integrity, including membrane structure; this may result in widespread and nonspecific perturbation of membrane function. A general protein reagent like Hg^{++}, for instance, is likely to produce a variety of membrane effects, both direct and indirect, as further illustrated in Fig. 5-2 (see section 5-4, below). Nonspecific perturbations may be produced by many chemical or physical agents. Examples are the consequence of peroxidation of membrane lipids, changes in membrane fluidity (see sections 4-3 and 5-4), swelling of cells with consequent membrane stretching (see Chapter 8), and redistribution of fixed electric charges resulting from binding of heavy metals (see section 6-3).

Extensive efforts have been made in the past to identify at the membrane the primary target site of many toxicants. Of course, interaction between membranes and nonlipophilic toxicants may represent, as indicated previously, no more than a transient and reversible step in toxicant uptake. This, by itself, need not lead directly to membrane malfunction. Mediated transport of toxicants across membranes is well established (see Chapter 6). It may consist of facilitation of cellular uptake of an agent, in which case it increases exposure of intracellular targets in the cytoplasm or associated with organelles such as nuclei or mitochondria. Alternatively, membranes may actively "export" or extrude toxicants from cells (see Chapter 7), thereby protecting the cells from the agent. In addition, membrane transport may permit selected toxicants to cross structures like the blood–brain barrier. In none of these instances do the membranes themselves necessarily show signs of malfunction.

Just as many examples may be cited of toxic effects believed to be directly exerted at the membrane, so in numerous other instances membrane lesions are well known to result indirectly from primary toxic reactions not associated with membranes. The renal salt–losing syndrome in adrenal insufficiency, for instance, represents a secondary result of interference with normal aldosterone production and secretion. The primary effect here is exerted on the adrenals; the consequent absence of adequate concentrations of aldosterone, in turn, secondarily depresses sodium reabsorption from the distal renal tubule. Membrane function also may be inhibited indirectly by toxicant action elsewhere in the cell itself. A classic example of such indirect (and in this case nonspecific) inhibition is the depression of active solute transport across the membrane in the presence of inhibitors of oxidative phosphorylation in mitochondria such as dinitrophenol, azide, or cyanide.

Subsequent to its cellular uptake, a toxicant may react with an intracellular primary target such as, for instance, oxidative enzymes in the mitochondria. This results in decreased production of ATP. The consequent depression of active transport at the cell membrane is independent of whether the energy is normally applied directly to support transport, as in the case of Na^+ extrusion mediated by ouabain-sensitive Na,K-ATPase, or whether it is spent to maintain appropriate transmembrane activity gradients of cotransported solutes. Such gradients are coupled to the uphill transport of other solutes, as for instance the proximal tubular reabsorption of glucose or amino acids, both of which are coupled to the simultaneous movement of Na^+ down its activity gradient into the cells. Finally, the nonhydrolytic and nonphosphorylating, or so-called purinergic, action of ATP in channel gating is disrupted by an inadequate supply of ATP (see sections 4-6, 5-5, and 7-3).

The result of metabolic inhibition recalls the effects of cold temperature on membrane-solute transport (see also Chapter 8). Certainly, mammalian cells cannot maintain homeostasis at 4°C. Temperature dependence of, for example, solute movement across a membrane, however, should not be equated automatically to the involvement of active transport. Indeed, many passive diffusion phenomena are characterized by Q_{10} values significantly greater than 1.0, as are intrinsic properties of media and membranes such as viscosity and fluidity. The viscosity of water, for instance, varies with temperature with a Q_{10} of 1.2. It follows that the suggestion of a contribution of active transport to renal uptake of Hg (Endo, Sakata, and Shaikh, 1995) or to Cd uptake by macrophages (Hart, 1978), largely on the basis of Q_{10} values that can be calculated not to have exceeded about 1.4, is entirely premature (see section 6-2).

5-3 SEPARATION OF DIRECT FROM INDIRECT LESIONS

The distinction between direct and indirect membrane effects, as pointed out previously, cannot always be drawn clearly. Nomiyama and Nomiyama (1994), for instance, suggested that the nephrotoxic action of Cd can be explained by malfunction at the proximal tubular brush border, independent of the concentration of the metal within the cell. If the reactive Cd originates from lysosomal breakdown of Cd-metallothionein, however, the primary site of action of the metal still may lie within the cell. One can argue, for instance, that the liberated metal primarily damages lysosomal membranes. In that case, the general cytotoxic effect reflected in the inhibition of plasma-membrane processes can result indirectly from the leakage of lysosomal enzymes into the cytoplasm. Another instance in which the distinction between direct and indirect membrane effects of Cd is similarly difficult was described by Jungwirth, Paulmichl, and Lang (1990): The metal was observed to enhance potassium conductance in cultured renal epithelioid (MDCK) cells. Such membrane malfunction of course may result from intracellular reactions.

A further example of the difficulty that may be encountered in attempts to identify direct effects on membranes is provided by the work of Shopicki et al. (1996), who concluded that a direct effect of gentamycin explains its inhibition of dipeptide transport at the brush border in rabbit kidneys. The drug did not change the affinity (Km) of the carrier for its substrate but decreased V_{max}. One also could speculate, however, that the inhibition results from interference with the proton gradient required for the dipeptide transport, rather than from a direct effect exerted on the dipeptide carrier.

Some toxicants may exert multiple actions on membrane function, both direct and indirect. This is well illustrated by the results of Sellinger, Ballatori, and Boyer (1991) on the mercury inhibition of alanine uptake by plasma membrane vesicles of the skate liver; the extent of this inhibition is seen in Table 5-1 by comparison of the control (Ia) and the Hg-treated (Ib) vesicles: 0.1 mM Hg depressed alanine uptake by 70%.

Note by comparison of Ia (control) and Ic (high K) that alanine uptake was severely depressed on replacement of extravesicular Na with K, that is, in absence of the Na gradient to which active alanine transport is coupled. Although prolonged exposure to

TABLE 5-1. *Effects of Hg^{++} on ^{3}H-alanine uptake by membrane vesicles of skate hepatocytes (Based on Sellinger, Ballatori, and Boyer, 1991)*

Incubation medium	Na gradient	Hg, μM	Alanine uptake, nmol/mg·min
I. No preloading			
a. Control	+	0	0.36
b. Control plus Hg	+	100	0.11
c. High K	−	0	0.10
d. Control plus Hg	−	100	0.09
II. Preloaded with Na and unlabelled alanine			
a. Control	−	0	0.20
b. Control plus Hg	−	75	0.10

Data from Sellinger et al. (1991).

Hg, in all likelihood, would abolish the Na gradient and thus depress alanine accumulation, the short exposure (1 minute) in the present experiments did not significantly alter the Na gradient. It follows that the metal inhibition seen in Ib must have resulted from some other mechanism, possibly based on a direct reaction of the toxicant with the membrane transporter.

A similar conclusion can be drawn from experiments IIa and IIb in Table 5-1. Here, the vesicles had been preloaded with Na and unlabeled alanine before labeled amino acid was added to the medium. Although the absence of the sodium gradient excludes the possibility of active transport of the tracer, the Hg sensitivity of tracer uptake still points to a possible direct depression of the alanine carrier. In other words, tracer uptake under these conditions presumably results from exchange diffusion mediated by a Hg-sensitive carrier.

The conclusion that mercury can inhibit alanine transport both by abolishing the Na gradient and by reacting with perhaps the transporter itself leaves unresolved the question of why Hg exerted little effect on net alanine uptake when the Na gradient was abolished by replacing Na with K, as in experiment Id. However, clearly the metal does inhibit alanine transport by more than one mechanism. This is further supported by a decrease in the intravesicular volume at somewhat higher Hg concentrations; shrinkage of the vesicles, in turn, reduces the intravesicular concentration of the amino acid per mg protein at equilibrium. Effects of Hg on water fluxes have been described by various authors, as in the studies of Grosso and De Sousa (1993) on apical water channels in toad skin and on the inhibition of aquaporins (see section 4-4). The metal also generally causes cells to swell, in large part probably through its inhibition of Na extrusion (see Chapter 8).

A similarly broad spectrum of inhibitory effects on membrane function, in all probability including both direct and indirect actions, was reviewed by Atchison and Hare (1994) for methylmercury. This compound increases the permeability of the neuronal plasma membrane to Ca, blocks voltage-dependent Ca and Na channels, and dissipates the transmembrane potential of isolated mitochondria. Clearly, early cellular effects of exposure to methylmercury are quite diverse, and the lesions induced may well be caused by different mechanisms.

A final example of multiple cellular effects of a toxicant, as yet not clearly identified as either direct or indirect, is the observation that both apical uptake and basolateral extrusion of amino acids reabsorbed in the proximal tubule of the rabbit kidney are depressed by cadmium (Foulkes and Blanck, 1991); the additional fact that the downhill basolateral extrusion of the amino acid analogue cycloleucine is as sensitive to inhibition by Cd as is its active basolateral uptake suggests a possibly indirect action of the metal on function at that membrane. The effect might be related, for instance, to cell swelling (see Chapter 8) and consequent alterations in membrane stretching and surface geometry. Both these effects, as further discussed in the next section, may lead to nonspecific changes in membrane function. An alternative explanation for the inhibition of the two opposing fluxes is a direct action of Cd on an exchange carrier, similar to the action of Hg on alanine transport described in Table 5-1.

5-4 MEMBRANES AS GENERAL (NONSPECIFIC)
PRIMARY TARGETS

Whether acting directly or indirectly on membranes, a toxic agent may well alter general membrane properties in such a way that many special functions become impaired. This section deals with direct but nonspecific agents interacting with the membrane.

An example of such direct and nonspecific membrane action is that of detergents that increase the general permeability of cell membranes. The hemolytic action of saponin, for instance, is well known. The effect of detergents on membrane function in other cells is illustrated here by the action of n-butanol. This compound can stimulate significantly membrane fluidity and passive membrane permeability. Thus, Bevan and Foulkes (1989), as documented in Fig. 2-3, reported that 1.5% (v/v) n-butanol greatly increases the initial rate of Cd uptake by brush border membrane vesicles from the rat jejunum. The alcohol also makes the membrane more permeable to the extracellular or extravesicular chelator EDTA, and thus renders all accumulated Cd accessible to the chelator.

Numerous other agents similarly have been reported to alter fluidity of membranes and thereby potentially to influence their function; one such toxicant is Pb (Zucker et al., 1988). Action of anesthetics on nerve conduction has been attributed to such mechanisms, as has been the effect of 17-β-estradiol on chondrocytes (Schwartz, Gates, and Nasalzky, 1996). Similarly, the neurotoxic action of organic chlorine compounds like the insecticide DDT is associated with changes in the Na and K permeability of the axolemma (Narahashi, 1996). The antiarrythmic drugs studied by Suwalsky et al. (1994) may perturb directly the bilayer structure of the membrane. A similar effect has been reported for the pesticide heptachlor (Suwalsky et al., 1997). Because these effects can be observed with synthetic phospholipid bilayers, they must result from direct interaction between toxicant and membrane. Although it is likely that the action of selected bacterial toxins on the permeability of tight junctions in intestinal epithelial cells, already referred to in section 4-6 (see also review by Schneeberger and Lynch, 1992), also represents a direct effect on membrane structure, an indirect action cannot be entirely excluded.

A generalized toxic action of phospholipase affecting membrane integrity has been described by Nagahama, Michiue, and Sakurai (1996). The enzyme activity was associated with the alpha-toxin of *Clostridium perfringens*, which bound to choline-containing phospholipids in liposomes reconstituted in presence of carboxyfluorescin. Over a range of 1.5 to 15.0 ng/mL, the toxin-stimulated leakage of the dye from liposomes in presence of Ca or other divalent metals known for their essential role in the enzymatic activity of the toxin. Release of carboxyfluorescin therefore presumably represents a consequence of the hydrolysis of membrane phospholipids. These experiments are obviously relevant to the action of the toxin on intact cell membranes.

Another instance in which potentially far-reaching and nonspecific effects on membrane function may be initiated directly at the membrane are changes in the distribution of electric charges that accompany, for instance, the binding of multivalent metal

TABLE 5-2. *Delayed metal inhibition of intestinal amino acid transport*

| | Zero Delay | | 1-h Delay | |
	k, %/min	Inhibition, %	k, %/min	Inhibition, %
Control	3.1	—	—	—
+0.5 mM HgCl$_2$	1.2	61	—	—
Control	2.0	—	1.0	—
+0.5 mM CdCl$_2$	1.9	5	0.5	50

Results from representative rats are shown; jejunal segments were perfused in situ with saline containing 0.5 mM cycloleucine. The zero-delay animals were then perfused for 1 minute with metal saline (preexposure) and cycloleucine uptake determined in presence of metals. Alternatively, animals were preexposed to saline (controls) or Cd, the perfusate washed out, and the animals maintained under anesthesia for an additional hour prior to uptake measurements. k, first-order rate constant determined over 16 minutes.
Data from Foulkes, (1991a).

cations. To the extent that such binding alters the transmembrane potential, it could influence the activity of voltage-gated solute channels important in the mediation of Ca, K, and other ion fluxes.

As further considered in section 6-3, neutralization of fixed anionic charges, at least in enteric epithelium, depresses the ability of the cells to take up cationic metals; at the same time it increases uptake of mercury present as anionic polychloride complex (Foulkes and Bergman, 1993). Binding of the metals to the ouside of the cell membrane (step 1A of metal absorption, see section 6-3) is a relatively rapid process, with an apparent half time of less than 15 seconds, the minimum time resolution of such measurements. No effect of temperature can be observed (Foulkes, 1991b). The subsequent transfer of bound metal into the cell (step 1B) varies with temperature ($Q_{10} \sim 1.4$), probably reflecting changes in membrane fluidity.

Although the rapid binding of heavy metals to the cell membrane may lead to immediate malfunction, other membrane processes are affected much more slowly. An example is the inhibition of amino acid absorption from the rat jejunum by Cd and Hg. This is documented in Table 5-2 by results obtained in a study of cycloleucine absorption from rat jejunum perfused in situ. Note that the effects of Cd became overt only after a significant initiation delay. During a 1-hour delay the control value for the exponential rate constant of cycloleucine absorption fell by 50% of its initial value, from 2.0% to 1.0% per minute; at that time point, however, the preexposure to Cd became reflected in an even greater fall in amino acid uptake (inhibition compared to 1-hour control was 50%). No such delay attached to the action of Hg.

The fact that, unlike Cd, Hg in these studies exerted an immediate inhibition supports the view that this metal acts directly although presumably nonspecifically on the membrane. This reaction may involve sulfhydryl and other groups on different proteins, including the transporter molecule itself. To explain the delayed action of Cd one might assume that the metal in this case acts indirectly on membrane transport. It is worth recalling, however, the earlier reference (see section 3-7) to the delayed Cd inhibition of alanine accumulation by purified renal membrane vesicles (Bevan et al., 1989). Because in such a preparation, inhibition can result only from direct actions on the membrane, the initiation delay observed in the intact tissue arises perhaps from restricted access of the metal to the transport site. The target of Cd in this system, for instance, might be buried deep in the membrane, or it actually might be located on the intravesicular side of the membrane. Alternatively, the primary effect of the metal may be related to a slowly developing reaction such as the peroxidative changes discussed subsequently. It must be mentioned that other workers found only short delays in the onset of inhibition of galactose absorption in rat jejunum by Hg (Rodriguez-Yoldi, Lluch, and Ponz, 1987) and Cd (Rodriguez-Yoldi and Ponz, 1987). Conceivably, metal binding groups involved in sugar transport are more easily accessible to Cd than those responsible for amino acid transport.

Evidence has been obtained supporting a binding of Cd both to the outside of the membrane and to less accessible sites deep in the membrane or on its innner aspect (see sections 4-3 and 6-3). There is first the observation that Cd accumulation in membrane vesicles is not affected by osmotic swelling or shrinkage, indicating that the metal in the vesicles is not free in solution but is bound to the membrane. Secondly, the bound metal consists of an EDTA-sensitive portion and a second pool accessible to the chelator only after destruction of the vesicles by, for example, freezing and thawing or the action of detergents. The reactions of metals with membranes are further discussed here and in sections 5-5 and 6-3.

The high affinity of heavy metals for a number of membrane constituents has been referred to in earlier chapters. Mercury, for instance, beyond its well-known binding to sulfhydryl residues in proteins, also reacts with hydroxy, carboxy, phosphate, imidazole, amino, and other groups. The interaction between methyl mercury and phospholipids is described by Girault, Boudou, and Dufourc (1997). The damaging effects of Cd on cell–cell junctions (Prozialeck and Niewenhuis, 1991) similarly may be nonspecific. Such an action of Cd also has been seen by Rossi et al. (1996) who reported that the metal increases the permeability of intercellular junctions in a cell line derived from human colon cancer (Caco-2 cells).

It is clear that, as illustrated for Hg in Table 5-1, metals can act as broad inhibitors of membrane function. The high general affinity of metals specifically for proteins also is the reason why metal sensitivity of a purified enzyme in vitro may possess little toxicologic significance (see section 3-1). Nevertheless, it is likely that factors such as the accessibility of specific sulfhydryl groups may render some enzymes more sensitive than others to inhibition by low metal concentrations. The action of metals on membrane function, therefore, although relatively general, nevertheless may exhibit limited specificity. To that extent, and like the at-times uncertain distinction between direct and indirect toxic effects on membrane function, the difference between general and specific effects of toxicants on membranes is not always clear.

Direct but nonspecific toxic effects on membrane function have been observed also with agents other than metals. For instance, Udden and Patton (1994) reported that butoxyacetic acid reduces the deformability of rat red cells. Cardiac β-adrenergic receptor function is altered by changes in the physicochemical properties of membranes in

cirrhotic rats (Zengchua, Meddings, and Lee, 1994). Like other lipids in biologic structures, those forming part of cell membranes are subject to nonspecific peroxidative
attack by various agents. The cytotoxic effects of peroxidation are well illustrated by
the work of Marin, Rodriguez, and Proverbio (1992). These investigators reported an
inverse relationship between lipid peroxidation and the activity of ouabain-insensitive
Na-ATPase in plasma membranes from rat kidney cortex. The suggestion was made
that the inhibition reflects the effects of changed membrane fluidity.

A number of other instances have been described in which lipid peroxidation interferes with membrane function. For example, Jourd'Heuil, Vaananen, and Meddings
(1993) demonstrated consequences of oxidant action on short-term glucose uptake
by membrane vesicles from the intestinal brush border. As measured by fluorescence
parameters, treatment with $FeSO_4$–ascorbic acid produced a progressive increase in
membrane rigidity, that is, loss of membrane fluidity. These changes were accompanied by appearance of malondialdehyde, a byproduct of lipid peroxidation. Simultaneously, the V_{max} value of glucose uptake decreased from 5.4 to 1.0 nmol·min^{-1}·(mg
protein)$^{-1}$. Restoration of membrane fluidity with a fluidizing agent (the octanoate
derivative A_2C) reversed the inhibition of glucose transport by about one third.

Little effect of high concentrations of L-alanine was seen in these studies on the
initial velocity of glucose uptake by voltage-clamped vesicles or on its depression
by peroxidation. The rationale for this control experiment was that the toxic effect
of peroxidation might be mediated indirectly by abolition of the sodium gradient;
this was expected to be abolished by the cotransport of the amino acid and sodium.
The interpretation offered instead for these results proposed that the inhibition of
glucose uptake by the oxidative stress is exerted in part by a direct action of reactive
oxygen species or other products of peroxidation on the transporter itself, and in part
indirectly by alterations of membrane integrity.

Production of malondialdehyde in the presence of such heavy metals as Cd or Pb
has been reported repeatedly in various systems. Lipid peroxidation was proposed
as the mechanism whereby Pb alters the properties of erythrocyte membranes (Jehan
and Motlag, 1995). Cohen et al. (1996) found that ozone inhibits cell membrane–
related and receptor-dependent functions such as the interaction of macrophages with
interferon; again, peroxidative changes were invoked to explain this effect. A role of
lipid peroxidation in the cytotoxic actions of Cd repeatedly has been suggested by
various authors (see, e.g., Ossola and Tomaro, 1995). The delayed onset of Cd effects
on membrane function discussed previously may be related to such a reaction.

The fact that the activity of membrane-associated enzymes may be influenced by
changes in their lipid microenvironment was emphasized by Broadway and Saggerson (1997). These workers studied the competitive malonyl-Coenzyme A (CoA)
inhibition of partially purified carnitine acyltransferase incorporated into liposomes.
If the liposomes were prepared from microsomal lipids of fed rats, greater inhibition
was observed than with lipids derived from starved rats. It follows that under these
conditions changes in membrane lipid composition rather than in the enzyme itself
determined enzyme activity.

Additional examples of the effects of toxic agents on the integrity of membrane
structures are illustrated by changes in the permeability of intercellular tight junctions

on exposure to certain natural toxins and to oxidants. The actions of some bacterial toxins on the barrier function of intestinal epithelium are reviewed in section 3-6. Similarly, exposure to oxidants also has been shown to increase the permeability of tight junctions in intestinal epithelial cells (Schneeberger and Lynch, 1992).

5-5 SPECIFIC ACTIONS DIRECTLY EXERTED ON MEMBRANES

Unlike the general reactions discussed in the preceding section, some toxic effects on membrane function are relatively specific for a carrier system, enzyme, or receptor. Section 4-2 refers to such direct membrane effects in reference to the activity of membrane-bound enzymes (ectoenzymes). Kim et al. (1997) reported that nicotinamide adenine dinucleotide oxidase in plasma membranes of HeLa cells is inhibited by impermeant conjugates of the antitumor compound sulfonylurea. Similar cell-surface sites of action had been described earlier for adriamycin and other anthracyclines covalently linked to polymers in order to minimize or prevent their cellular uptake. The action of the cardiac glycoside ouabain on the activity of Na,K-ATPase constitutes another example of direct and specific inhibition of a membrane enzyme. As pointed out in section 5-2, however, inhibition of ATPase by the cardiac glycoside also can lead indirectly to depression of Na-coupled transport of various organic solutes across the membrane.

Note that competitive transport inhibitors like Probenecid, the synthetic inhibitor of organic anion transport in the kidney, or, for instance, the action of diuretics on sodium reabsorption, are excluded from this discussion. This is justified by the fact that they represent specific and generally fully reversible pharmacologic antagonists rather than toxicants.

A useful technique for the demonstration of direct toxic actions on membrane function is the use of purified membrane vesicles or reconstituted liposomes (see section 3-7). As pointed out, however, extrapolation from such preparations to the intact animal may be difficult and does not help identify necessarily the primary target of a toxicant in vivo. Thus, as repeatedly mentioned, the mercury inhibition of a purified enzyme incorporated into liposomes gives little indication of the primary target of the metal in vivo. In addition, as for instance in the case of glucose transport, it is important to exclude possible toxic effects on the closely coupled sodium gradient before the action of a toxicant can be attributed to its direct reaction with the glucose transporter.

Prior to the recognition of signal transduction and multiple messengers, the actions of ouabain and other agents at cell membranes were cited in support of the attractive hypothesis that many, if not most, physiologic regulators, hormones, and therapeutic drugs exert their specific primary action directly on active membrane sites (see Chapter 1). The concept of membrane lesions responsible for the toxic effects of xenobiotics had been clearly formulated by Rothstein (1959), in particular reference to heavy metals. Figure 5-1 provides some of the evidence that originally helped define glucose transport across the cell membrane as one of the immediate targets for acute effects of mercury. A direct effect of Hg on a membrane transporter also was

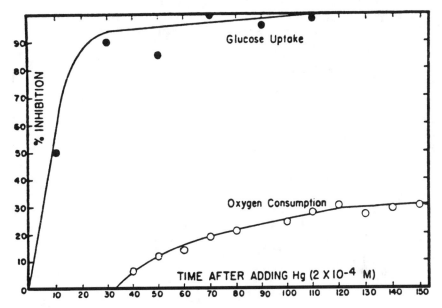

FIG. 5-1. Action of Hg on oxygen and glucose uptake by rat diaphragm. (From Rothstein, 1959; with permission.)

suggested by the results shown in Table 5-1 on mercury inhibition of alanine uptake by membrane vesicles.

Figure 5-1 shows that the membrane is likely to be a major, although not necessarily the only, target of 0.2 mM HgCl$_2$. This admittedly high concentration of Hg depresses both glucose uptake and respiration of rat diaphragm. Note, however, that glucose uptake is inhibited relatively rapidly, at a time when respiration still remains unchanged. A similar relatively rapid onset of mercury inhibition of amino acid absorption from the rat jejunum is documented in Table 5-2. The results are compatible with the hypothesis of a direct effect of the metal on membrane function, coupled with its slow penetration into the cells and consequently delayed effect on oxidative metabolism in mitochondria. The validity of such an interpretation is put into some doubt, however, by the conceptually similar observation that azide depresses potassium uptake by yeast before inhibiting respiration (Foulkes, 1956). The hypothesis here requires the assumption of a direct effect of the inhibitor on K fluxes. The possible nature of such a reaction is not clear.

The mercury inhibition of glucose uptake illustrated in Fig. 5-2 could be reversed with cysteine; no such reversal was observed for the depressed respiration. Such findings offer further support for a primary effect of Hg on glucose transport at an accessible site on the membrane. Suspected direct effects of mercury (and other metals) at synaptic membranes also have been described, as reviewed by Verity (1996). Another example of presumably direct Hg interference with membrane function is the reaction of the metal with sulfhydryl groups essential for glucose transport at

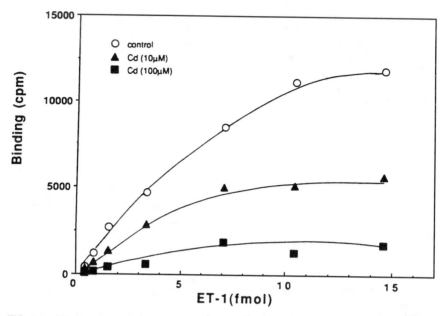

FIG. 5-2. Cd inhibition of endothelin binding to its receptor. Abscissa: fmol of labelled endothelin-1 in 50 μL reaction mixture; ordinate: radioactivity bound to purified receptor. (From Wada et al., 1991; with kind permission of Elsevier Science-NL, Sara Burgerhartstraat 25, 1055 KV, Amsterdam, The Netherlands.)

the erythrocyte membrane (Vansteveninck, Weed, and Rothstein, 1965). The ability of the metal to weaken the anchoring of Na,K-ATPase in the membrane (Imesch, Moosmayer, and Anner, 1992) also is likely to represent a primary action on the membrane. It must be pointed out that much of this work, however, employed relatively high metal concentrations, and it would have been surprising if the metal had not reacted freely with many membrane proteins; the full toxicologic significance of the findings therefore remains uncertain.

An informative summary of studies on membranes as targets especially of heavy metals and chlorinated insecticides was presented by Kinter and Pritchard (1977). Effects of DDT on nerve membrane conductance, for instance, are well documented. They may be quite complex, involving interference with such functions as gating of ion channels and activity of membrane Na,K-ATPase. Quite generally these compounds exert their toxic effects through alterations in solute handling by the membrane; as a result they can affect nerve conductance, muscle excitability, and many other cell processes, including maintenance of cell volumes and osmoregulation of body fluids. Chapter 8 further discusses the role of cell membranes in volume homeostasis, and its sensitivity to toxic interference.

Many natural toxins have been suspected of acting primarily at the cell membrane. A number of these now are known specifically and directly to inhibit inorganic ion fluxes across cell membranes. They include the cardiac glycoside ouabain, the

TABLE 5-3. *Some natural toxins directly reacting with sodium channels*

Name	Source
Tetrodotoxin	Puffer fish
Saxitoxin	Dinoflagellates
Brevetoxins	Dinoflagellates
Ciguatoxins	Dinoflagellates
Conotoxins	Sea snail
Batrachotoxin	Frog
Alpha and beta toxins	Scorpion
Sea anemone toxin	Marine plant

Data from Catterall (1992).

classic and specific inhibitor of Na,K-ATPase, the enzyme responsible for Na transport out of cells. Other toxins react with specific membrane receptors, as in the case of bungaratoxin from snake venom, which binds to acetylcholine receptors (see, e.g., Lozano, Yee, and Buchanan, 1994). Alpha-scorpion toxin and sea-anemone toxin have been shown to react with specific amino acids in the extracellular domain of mammalian Na channels (Rogers et al., 1996). Some of the natural toxins capable of interacting with Na channels are identified in Table 5-3.

A number of these natural toxins and channel modifyers have been isolated from a variety of biologic sources, ranging from microorganisms to vertebrates. Not only movement of Na may be affected; other solute channels and pumps also react with specific toxins and have been mentioned previously. Other examples are nifedipine, the plant alkaloid ryanodine, the spider venom agatoxin, and maitotoxin from dinoflagellates, all specific for calcium channels. Similarly, 4-aminopyridine, tetraethylammonium, the snake venom dendrototoxin, and charybdotoxin from scorpions interact with potassium channels. Anthracene-9-carboxylate acts as chloride channel antagonist (see, e.g., Yu et al., 1997). The evidence presented by Carrasquer and Schwartz (1996) points to a direct effect of the bee venom melittin on the H,K-ATPase (the proton pump) of frog gastric mucosa. Melittin also reacts with Na,K-ATPase (Cuppoletti and Abbott, 1990). The interaction between the wasp venom mastoparan and biomembranes is reviewed by Katsu et al. (1990).

Adams and Swanson (1994) summarized information on many toxins shown to affect solute channels in cell membranes of the nervous system. In many instances, these interactions involve direct effects on the solute channels, and the toxins thus can serve as molecular probes for these channels. In other cases, inhibition of a membrane transporter may result not from a direct action of the toxin on the protein but, for instance, from possible interference with the exocytotic processes (membrane traffic) bringing the transporter to the outer membrane. This may explain the action of the macrolide antibiotic bafilomycin A on activation of the H^+-ATPase-mediated proton extrusion from human neutrophils (Nanda et al., 1996).

The toxicologic significance of alterations in pump and channel activity obviously derives from the fact that cell function closely depends on cell composition and thereby to a significant extent on membrane permeability. To list two further instances of this dependence, reference may be made to the observations that neurotransmitter release in nerve endings is triggered by a rise in Ca^{++} levels, and that nerve or muscle excitability requires appropriate sodium and potassium activity gradients across the membrane. Any xenobiotic that affects, for instance, the normal changes in Na and K permeabilities of the nerve cell membrane during passage of an action potential, may thereby abolish nerve conduction. Among a great variety of compounds acting in this manner are the local anesthetics referred to elsewhere.

Many natural toxic compounds act on membranes in a less specific manner than do the channel toxins just described. Such actions may lead to increased general permeability of the membranes. An instance is the hemolytic effect reported for certain snake and spider venoms. This repeatedly has been shown to involve phospholipase action on membrane phospholipids, as is the case with the *Clostridium perfringens* toxin discussed in section 5-4. Indeed, more than 100 out of almost 300 known bacterial protein toxins have been reported to permeabilize cell membranes, in selected cases by opening large and poorly selective channels (Menestrina, 1994). Such an action is characteristic, for instance, of the α-toxin from *Staph. aureus* (Fussle et al., 1981), and of the effect of the Kanagawa hemolysin from *Vibrio hemolyticus* on the cation permeability of the erythrocyte membrane (Huntley and Hall, 1996).

Many natural hemolytic compounds other than the bacterial toxins have been described; one of these is the well-known plant product saponin. In general, any chemical possessing detergent properties, as do soaps, alcohols, and other classes of compounds, tends to destroy membrane integrity.

In general, the specific channel blockers react directly with the membrane; the action of others is more indirect. Thus, the presence of ATP is required for maintaining certain channels in an open state; function of the channel is thus indirectly and nonspecifically depressed by metabolic inhibitors. An example of such a channel is the multiple drug resistance channel discussed in Chapter 7. In this and similar cases the primary action of a channel blocker may consist of interference with ATP synthesis, with a resulting fall in intracellular or even extracellular ATP. This could explain in part the action of high concentrations of Hg on the ATP-requiring taurine channel in coelomocytes of marine worms (Preston et al., 1991). Lower concentrations of the metal, however, by themselves insufficient to affect intracellular ATP/ADP ratios, suffice to depress channel activity. The effect of the metal therefore may involve a direct interaction with the channel, such as seen in Table 5-1 for the amino acid transporter in membrane vesicles from skate liver.

Purified membrane preparations such as vesicles provide a useful tool for defining mechanisms of interaction of toxicants with membranes. If the toxicant is administered to the intact animal, before the preparation of vesicles, any transport lesion observed, of course, may have resulted indirectly from primary lesions elsewhere in the body. In contrast, effects observed in purified vesicles exposed to toxicants in vitro must represent actions exerted directly on the membrane, even though the toxicologic significance of such a finding for the intact animal may remain in doubt. It also must be remembered that a direct interaction between membranes and toxicants resulting in the loss of a specific function should not be equated with a direct effect on the particular transporter: A toxicant, for instance, could abolish the Na gradient required for glucose transport (see section 5-3).

A variety of effects on membrane function, involving direct interaction between toxicant and membrane, have been observed in this manner in vitro. For instance, the antineoplastic platinum compound cisplatin, whose therapeutic usefulness is limited by its nephrotoxicity, inhibits organic ion transport by membrane vesicles from rat renal cortex (Williams and Hottendorf, 1985). Verbost et al. (1988) described the competitive Cd inhibition of ATP-requiring Ca transport in membrane vesicles from fish gills. The primary effect here appears to be exerted on the Ca-binding sites of the enzyme; secondary effects in vivo may be related to the changes in cellular Ca metabolism. Cadmium inhibition of L-alanine uptake by brush border membrane vesicles of the flounder kidney and of Na-coupled glucose transport by brush border membrane vesicles of rat jejunum (see section 3-6) provide further examples of membrane sites of action of the metal.

Some of the difficulties of extrapolating from an isolated system to the intact organism are detailed in Chapter 3; they have not always been taken into account fully. The work of Oliveira, Rocha, and Sarkis (1994) can serve to reemphasize this important point. These investigators found that 10 μM $HgCl_2$ inhibits the ATPase activity in isolated synaptosomes from rat cerebral cortex; very high doses in vivo, in contrast, remained without effect on this enzyme. The neurotoxic effects of the metal in vivo therefore cannot be attributed to an action on synaptosomal ATPase. Presumably the negative finding in vivo at least in part can be explained by the low permeability of the blood-brain barrier to inorganic Hg. In any case, the finding further supports the critical conclusion that toxic effects of an agent on a specific enzyme or compound in vitro do not necessarily predict the nature of its primary target in vivo.

Additional examples of presumably direct effects of heavy metals on membrane function in vitro include, for instance, the effect of Pb on the activity of inositol polyphosphate receptors on cerebellar membranes (Vig et al., 1994). Abrami et al. (1996) observed that passage of glycerol through human water channels (aquaporins) expressed in oocytes is sensitive to inhibition by p-mercuribenzenesulfonate and Cu sulfate. Reference has already been made in sections 4-4 and 5-3 to the metal sensitivity of the aquaporins.

Effects of heavy metals on serotonin-induced currents in mouse neuroblastoma cells are attributed to their direct reaction with the serotonin receptor–ion channel complex (Uki and Narahashi, 1996). Giridhar, Rathninaveln, and Isom (1992) reported that cadmium inhibits the receptor for atrial natriuretic peptide. Activity of the ectoenzyme 5'-nucleotidase is inhibited by cadmium in a dose-dependent manner (Morselt et al., 1987). The metal in relatively high concentrations also interferes with binding of the peptide hormone endothelin to specific membrane receptors from human placenta (Wada et al., 1991). The studies on endothelin binding involved incubation of ^{125}I-labelled hormone (endothelin-1) with isolated receptor (a protein of 40 kD molecular weight). The inhibitory effect of Cd on the iodine binding in these studies presumably was exerted directly on the receptor or the hormone. Results of these experiments are illustrated in Fig. 5-2.

Note that Cd inhibition is dose-dependent; in presence of 3 fmol endothelin, 0.1 mM Cd inhibits binding of the agonist by over 90%, compared with less than 50% at 0.01 mM Cd. At either high or low Cd levels, however, the fractional inhibition is independent of substrate concentration. In other words, the metal here appears to react directly and noncompetitively with its target from the membrane. A similar independence of fractional inhibition of solute binding and uptake from the concentration of that solute is observed in studying the effects of polyvalent cations on jejunal Cd transport (see section 6-3). This similarly appears to represent a direct and noncompetitive inhibition of membrane function.

The full toxicologic significance of the reported Cd effect on endothelin receptors is not entirely clear. First there is the fact repeatedly mentioned in earlier chapters that demonstrating the inhibition of a purified receptor in vitro does not establish this receptor as the toxicologic target in vivo. Questions are also raised by the inability of Koschel, Meisner, and Tas (1995) to observe such an effect of Cd with endothelin receptors still associated with membranes; steric or other factors may influence the reaction between metal and receptor. The attenuation of endothelin-induced aortic strip contraction by Cd may be related to an action of the metal on the hormone-stimulated Ca influx into cells.

Note that the widespread effects of the metals reflect in part at least their high and nonspecific affinity for proteins, including membrane proteins. Although they do react directly with membranes, their toxic action in general is not specific. A direct and specific toxic membrane effect, in contrast, results from the well-known action of acetylcholine antagonists on receptors at the neuromuscular endplate. Transmission of action potentials by acetylcholine released from the nerve terminals is inhibited by, for example, tubocurarine, which binds with high affinity to the acetylcholine receptor on the postsynaptic membrane. In summary, many well-documented examples are cited in this section of specific and direct interactions between toxicants and cell membranes.

5-6 INDIRECT MEMBRANE LESIONS

Some 40 to 50 years ago there was strong support for the attractive hypothesis that cell membranes are the primary site of action of perhaps the majority of physiologic messengers (hormones), therapeutic drugs, and toxic agents. For instance, insulin was believed somehow to stimulate directly the glucose transport system. The hypothesis proved simplistic, however, and we now know that the primary action of insulin lies inside the cell. Indeed, recognition of intracellular signal transduction (second or third messengers), cytoplasmic and nuclear receptors, mitochondrial inhibitors, blockers of membrane traffic, and other factors clearly shows that many toxic effects, even though they ultimately may alter membrane function, are in the first place elicited inside the cell. Obvious examples are genomic toxicants and metabolic poisons like cyanide or dinitrophenol.

The metabolic inhibitors nonspecifically and indirectly abolish active solute transport across the membrane mostly by interrupting ATP synthesis and consequently the supply of metabolic energy. ATP also may play a nonhydrolytic or nonphosphorylating role in membrane function by determining the gating of certain solute channels. Such an apparently purinergic action of ATP was demonstrated, for instance, for the function of the drug-extruding glycoprotein Pgp discussed in Chapter 7 and for the taurine channel involved in cell volume homeostasis (see Chapter 8). The observations that exposure to a toxicant may stimulate synthesis of mRNA and that inhibitors of protein synthesis like oligomycin may prevent the cytotoxic effects of certain xenobiotics further emphasize the importance of intracellular sites of action.

Unlike the broad spectrum of effects on membrane function indirectly caused by these metabolic inhibitors, other indirect lesions may be relatively more specific. An example of such inhibition is found in the work of Clemens et al. (1996); these investigators reported that ethanol depresses the binding of asialoglycoprotein to its membrane receptor. The impairment is mediated by the action of alcohol oxidase, suggesting the possibility that the active agent is acetaldehyde. Similarly, an apparently indirect but specific action probably is involved in the effects of the pesticide lindane on gap junctions in uterine smooth muscle (Criswell and Loch-Caruso, 1995); this is mediated by an arachidonic acid–sensitive, cAMP-independent mechanism.

Indirect membrane effects of toxicants cannot be excluded even if a toxic agent is shown to react directly with membrane constituents. Indeed, the reaction of a xenobiotic at the cell membrane may represent only a transient step in its uptake and ultimate reaction with an intracellular receptor. Alternatively the reaction may serve to trigger a cascade of events involving multiple messengers. Such a series of reactions may indirectly lead to functional lesions at the membrane. For instance, binding of heavy metals to the membrane, as discussed above, can interfere directly with various processes. In part, however, the binding is transient and simply constitutes a first step in metal uptake (see section 6-3). In that case the metals, although directly reacting with the membrane, still can exert subsequently an indirect effect on membrane function.

It is worth repeating the caution expressed in section 3-7 that analysis of such indirect actions may require a relatively intact experimental model, and that these actions are not always amenable to study in, for example, isolated and purified membrane vesicles. Consider, for instance, the acute effect of Cd on basolateral amino acid uptake in renal cortex in vivo. If this results indirectly and presumably nonspecifically from actions on such characteristics as cell volume and thereby cell geometry, as suggested by Foulkes and Blanck (1991), little inhibition of peak amino acid accumulation in basolateral membrane vesicles might be expected.

The evidence for nonspecific Cd effects on basolateral solute transport was obtained by comparing the responses of two different solute transport processes to the metal in kidneys of intact rabbits (Foulkes and Blanck, 1991). In those experiments, steady-state concentrations of PAH and the amino acid analogue cycloleucine were determined in the cortex of nonfiltering (stop-flow) kidneys. Active basolateral influx of the two unrelated solutes was depressed in rabbits acutely poisoned with Cd. At the same time their steady-state tissue-to-plasma concentration ratios remained unchanged. This fact implies that passive leakage of PAH and cycloleucine from cells into interstitial fluid must be as sensitive to Cd as is their active uptake in the opposite direction. As pointed out, these findings support the interpretation that Cd acts indirectly on the transfer of the two solutes, and that the observed inhibition may result from a general and nonspecific change such as cell swelling or from damage to membrane integity resulting from, for example, lipid peroxidation. Alternatively, if one assumes that solute efflux is mediated by the same carrier system as is influx, the finding can be explained by a direct reaction of the metal with the solute carriers.

The action of toxicants on cell swelling, and on the control of cell volume in general, is further considered in Chapter 8. In the case of the Cd inhibition of basolateral solute fluxes under discussion here, any underlying changes in cell volume may result from the normal response of cells to metabolic inhibition. The connections among cytotoxicity, interference with energy metabolism, and the control of cell volume long have been recognized, as for instance in the swelling of cells exposed to cyanide, anoxia, or cold. General metabolic inhibition, however, is not the only mechanism whereby cellular volume control may be threatened. Thus, Wallin and Hartley-Asp (1993) listed some agents capable of inhibiting the assembly of microtubules; such inhibition may affect cell architecture including volume. Cadmium is one of the agents identified as potentially reacting with microtubules, but even at relatively very high concentrations (1 mM) the inhibition is only partial. Another indirect mechanism of changing membrane function primarily can involve, for instance, cell swelling as a result of changes in osmolyte fluxes and metabolism, with consequent stretching of the cell membrane. Stretch effects on the function of solute channels in the membrane have been well established (see section 3-7).

The observations on the time delay required before Cd exposure leads to overt membrane malfunction (see section 5-4) also are compatible with indirect actions of the metal. Assuming that the conclusions obtained on amino acid transport across the renal tubular epithelium (Foulkes, 1972) may be extrapolated to the jejunal epithelium, the primary transport process in intestinal amino acid absorption occurs at the brush border. Obviously the initiation delay in the perfused intestine in that case cannot be ascribed to the time required for the metal to reach a critical concentration at the brush border. Presumably the target site of Cd, if it is associated with the membrane, is therefore either not readily accessible from the extracellular fluid (see section 5-4), or the toxic action of Cd on membrane function is indirect and involves primary targets elsewhere in the cells. It is interesting here to refer to another instance of time-dependent Cd toxicity described in human T cells (El Azzouzi et al., 1994). Of course, the consequence of a direct interaction between a toxicant and the membrane also can be delayed if it results, for instance, from some slowly developing effect of oxidative stress. As discussed in section 5-4, a variety of toxic solutes, including heavy metals, can cause peroxidative changes in membrane lipids.

The contrast between the delayed action of Cd and the immediate effect of Hg on solute absorption in the intestine is emphasized in section 5-4. The apparently direct reaction of Hg with glucose transporters in yeast-cell membranes also may be recalled (see section 2-4). Clearly, a xenobiotic may exert multiple actions. Further, as discussed in section 5-3, identification of any one of these actions as a direct or indirect inhibition of membrane function may be difficult at times.

5-7 TOXIC ACTIONS ON SUBCELLULAR MEMBRANES

Although emphasis has been placed so far on malfunction at the plasma membrane separating the cytoplasm from the extracellular fluid, somewhat similar lesions also have been observed at subcellular membranes surrounding, for instance, nuclei, mitochondria, lysosomes, peroxisomes, and other subcellular organelles. The morphology of these membranes has been studied extensively under the electron microscope, and significant information is available on their permeability properties. Like outer cell membranes, these intracellular membranes are associated with a variety of enzymes, transporters, channels, and receptors, and are thus potentially sensitive to toxic inhibition of their function.

For instance, decreased supply of ATP lowers the ion permeability of isolated mouse liver nuclei (Assandri and Mazzanti, 1997). Inhibition of metabolism, on the other hand, causes swelling of mitochondria (see Chapter 8). Lesions at the mitochondrial membranes may lead to cytotoxic consequences by affecting the basic metabolic functions of these organelles. An instance is the cytotoxicity commonly observed resulting from inhibition of mitochondrial Ca sequestration, as further discussed subsequently. Damage to lysosomal or peroxisomal membranes can lead to leakage of normally sequestered enzymes into cytoplasm, with potentially serious consequences.

Some further examples of catalytic functions associated with subcellular membranes are given here; each of these functions is a potential target of inhibition by toxic agents. Pisoni and Thoene (1991) reported the presence of multiple carrier systems in lysosomal membranes mediating transport of amino acids, sugars, nucleosides, inorganic ions, and so forth. Nuclear membranes of eucaryotic cells were assigned active roles in the transport of proteins and RNA (Gorlich and Mattay, 1996). Maruyama, Shimada, and Taniguchi (1995) found Ca-activated K channels in the nuclear envelope isolated from single pancreatic acinar cells. Cyclooxygenase activity was observed in nuclear membranes of a human leukemia cell line (cited by Matsumoto, Morita, and Murota, 1994). Whatever the full functional significance of these findings, it is clearly likely that inhibition of subcellular membrane function may be as closely associated with cytotoxicity as is malfunction of the outer cell membrane.

Inhibition of subcellular membrane function by toxic agents is better documented in mitochondria than in the other organelles. For example, acylcarnitine serves as substrate of a specific transferase and in this capacity mediates transfer of fatty acid substrates across mitochondrial membranes. Xenobiotics like heavy metals or cephaloridine, known to interfere with carnitine metabolism, may exert some of their broad effects on cellular metabolism by inhibiting this mitochondrial transport function (Garst, 1995); activity of carnitine acyltransferase is depressed by malonyl-coenzyme A (Ghadiminejad and Saggerson, 1991). Zara et al. (1996) partially purified and reconstituted a tricarboxylate carrier from eel-liver mitochondria; transport of 0.1 mM citrate was 89% inhibited by 1 mM 1,2,3-benzenetricarboxylate. Such inhibition may interfere greatly with tricarboxylic-acid metabolism.

Other enzymes or processes directly associated with microsomal membranes also are subject to toxic inhibition. Thus, ferrous iron appears directly to depress Ca^{++} release from vesicles of cardiac sarcoplasmic reticulum (Kim, Giri, and Pessah, 1995). This is believed to involve a competitive interaction between Ca^{++} and Fe^{++} at the ryanodine receptors. Heavy metal ions and sulfhydryl-oxidizing agents like anthraquinones also modify the gating behavior of ryanodine-sensitive Ca^{++} channels that had been purified from sarcoplasmic reticulum and incorporated into artificial lipid membranes (Abramson et al., 1988; Abramson and Salama, 1989).

These observations all emphasize the toxicologic significance of intracellular membranes as sites of inhibitory action. As is the case with the outer cell membrane, such toxic effects may represent direct or indirect results of exposure. In either case, the integrity of the organelle may be severely affected. This is seen especially in the area of volume regulation (see Chapter 8). Indeed, like intact cells, mitochondria control their internal composition and thereby their volume (Tedeschi and Harris, 1955); the solute permeability of the membrane is a major factor in this process. Mitochondria are capable also of compensatory volume regulation, a process linked to metabolism and to the solute permeability of the membrane.

The channels facilitating transmembrane fluxes in mitochondria, as in the cell membrane, are sensitive to various inhibitors. For instance, the anion channel in the inner mitochondrial membrane is inhibited by the mercury compound mersalyl (Beavis, 1992). This obviously is not a specific inhibition, as a number of other carrier proteins active at that membrane are similarly affected. The anion channel further is sensitive to tributyl tin, with a concentration of 1 nmol/mg mitochondria exerting strong inhibition. This is, however, a relatively high concentration (1 mM on a volume basis, assuming a specific gravity of 1.0); the specificity of such an effect is uncertain.

Actively metabolizing mitochondria maintain a high transmembrane potential and establish significant concentration gradients of organic and inorganic electrolytes across the limiting membrane (see, for example, MacFarlane and Spencer, 1953; Aw, Andersson, and Jones, 1987). The sensitivity of the membrane potential to toxic xenobiotics has been reported repeatedly, as for instance by Yuan and Acosta (1996). These authors found that the potential is depressed in rat myocardiocytes by a noncytotoxic concentration of cocaine (0.5 mM).

A variety of cytotoxic agents exert their actions on mitochondria by nonspecifically increasing the permeability of the inner membranes. This so-called permeability transition (Gunter et al., 1994) is accompanied by opening of a complex, cyclosporine-sensitive channel whose activity is modulated by changes in the membrane potential (Scorrano, Petronilli, and Bernardi, 1997). The transition is induced by inhibition of the electron transport chain (Chavez et al., 1997); it is activated by antimycin A. Other xenobiotics that have been shown to elicit such transitions include oxidants, metals, and electrophiles capable of forming thioether adducts (Henry and Wallace, 1995).

The permeability transition is characterized by Ca-induced membrane depolarization, loss of ion and volume regulation, and inhibition of oxidative phosphorylation; different xenobiotics may produce these effects by different mechanisms.

Use of isolated organelles as model systems for the study of cytotoxicity, as with any other isolated preparation, requires the implicit assumption that they behave in vitro as they do in their physiologic environment in the living cell (see Chapter 3); this assumption is not readily tested. In spite of such uncertainty, use of purified preparations has furnished important information. The broad effects of the immuno-suppressive drug cyclosporin on membrane function in subcellular organelles provide a good illustration of the potential for cytotoxicity at that level (Whiting, 1993). This agent influences solute permeability of mitochondria isolated from various tissues; at therapeutic concentrations the drug impairs Ca-induced swelling (Strzelecki et al., 1988). Nuclear membranes also are affected and show significant lobulation (Simons et al., 1986). The broad susceptibility of mitochondrial membranes to toxic agents is further illustrated by the actions of some antibiotic and antineoplastic compounds. For instance, cisplatin influences mitochondrial permeability (Kameyama and Gemba, 1991); doxorubicin (adriamycin R) activates mitochondrial Ca channels (Solem and Wallace, 1993).

Many of these toxic effects have been studied in isolated, subcellular organelles, but significant information on subcellular membrane function also has been gained from intact cells. Thus, interference with normal calcium metabolism, such as release of Ca from intracellular stores or inhibition of its sequestration by subcellular organelles, or changes in Ca movement across the cell membrane, may be involved in the effects of many toxicants. It is noteworthy that Ca channel blockers sometimes can prevent toxic manifestations of exposure. For example, they partially protect the liver against carbon tetrachloride (Romero et al., 1994). The general importance of alterations in Ca homeostasis as an immediate cause of cytotoxicity has been documented extensively, as for instance in the review by Viarengo (1994).

Studies on mechanisms responsible for toxic lesions at intracellular membranes in living cells, and on their significance to the intact organism, are complicated by the uncertain concentrations of toxicants at their sites of action. Integrated areas under the curve describing changes in the plasma concentration of a toxic agent with time do not necessarily provide much information on concentration profiles at, for example, mitochondrial membranes.

The problem of establishing the cytoplasmic concentration of a toxicant in the living cell at a given time is complicated by the fact that total cellular accumulation of a xenobiotic is not very informative in absence of knowledge about its intracellular volumes of distribution. Cytoplasmic concentrations, of course, cannot be calculated from total cell accumulation unless the volume of cytoplasm is known, and the uptake of the xenobiotics by organelles can be estimated. In other words, it is difficult to obtain an accurate measure in intact tissues of the concentration of a toxicant at a potential intracellular target site such as the mitochondrial membrane.

Attempts have been made with both morphometric and functional procedures to determine the volumes of cytoplasm and organelles in living cells. One technique employed 3-O-methylglucose (3OM) as volume probe (Foulkes and Blanck, 1994). Many cells, including lymphocytes freshly isolated from rat blood, appear to be freely permeable to this nonmetabolizable sugar analogue but do not actively accumulate it; cytoplasmic concentration of the sugar at steady state therefore should equal its extracellular concentration. On that basis, the volume of distribution of 3OM (V_{3OM}) in the lymphocytes was found to include only 40% of total cell water; the difference, by definition, must represent the volume of 3OM-impermeable subcellular structures. It follows that intracellular V_{3OM} can provide a maximum estimate of the cytoplasmic volume in cells like these lymphocytes, provided that they are passively permeable to the sugar. In that case, the maximum cytoplasmic concentration of a xenobiotic equals its total cellular content divided by V_{3OM}. To the extent that the xenobiotic does bind to organelles or cytoplasmic constituents, its free concentration in the cytoplasm will be lower than the calculated value.

Incomplete mixing of sugar probes with total cell water, indicating presence of nonpermeable organelles, has been observed in other tissues and especially in renal cortex (Silverman, Aganon, and Chinard, 1970; Ullrich and Papavasiliou, 1985). In contrast, Latzkowits et al. (1993) reported that in cultured astrocytes V_{3OM} is approximately coextensive with cell water. Such a finding can be interpreted as indicating that organelles occupy only an insignificant fraction of the cell volume, or that the organelles in these cells are permeable to the sugar analogue, or even that some sugar is actively accumulated in cytoplasm.

In fresh rat lymphocytes, as indicated above, V_{3OM} occupies only about 40% of cell water (Foulkes and Blanck, 1994). In some preliminary studies on intracellular effects of mercury, the cells were exposed for 10 minutes to 0.1 mM $HgCl_2$. As expected from the known effects of the metal on cell permeability and metabolism (see Chapter 8), this exposure led to extensive swelling of the cells, total cell water more than doubling. These results are shown in Table 5-4.

Note that the swelling of cells described in Table 5-4 essentially can be accounted for completely by the increase in the cytoplasmic volume (V_{3OM}), while little change occurred in the portion of cell water not accessible to 3OM, i.e., the volume of 3OM-impermeable organelles. Whether because of the protective action of the cell membrane (see section 2-4), or for other reasons, the 3OM-impermeable membranes of

TABLE 5-4. *Intracellular water compartments in rat lymphocytes*

	Control	Exposed to Hg	Change
Cell water (V_{H_2O})	3.2	7.8	4.6
Cytoplasm (V_{3OM})	1.4	5.7	4.3
Organelles ($V_{H_2O}-V_{3OM}$)	1.8	2.1	0.3

Cells freshly isolated from plasma were preincubated for 10 minutes at 37°C with or without 0.1 mM $HgCl_2$. Volumes are expressed as mL/g protein.
Data from Foulkes (unpublished experiments). For procedures, see Foulkes and Blanck (1994).

subcellular organelles under these conditions remained largely unaffected by the short exposure of the intact cell to Hg.

The toxicologic significance of these findings is difficult to interpret. A critical problem here, as already emphasized, is that the metal concentration reached at the intracellular membranes in these experiments is not known. It is inappropriate, therefore, to conclude that the internal membranes are less Hg-sensitive than the outer cell membrane. Conceivably also this illustration of an apparent insensitivity of subcellular membranes to mercury could reflect the requirement for more prolonged exposure before the toxicant reaches a critical concentration in the cell. In the work of Guo et al. (1997), as presented in an abstract, 30 minutes' exposure of human T cells to 36 μM $HgCl_2$ reduced mitochondrial transmembrane potentials; it is not stated whether shorter exposure periods would have sufficed to cause similar changes. In general, the uncertainties about toxicant distribution within cells increase the difficulty of studying in the living cell the mechanisms whereby toxicants inhibit the function of subcellular membranes.

5-8 MEMBRANES IN IMMUNOTOXICOLOGY

A wide variety of chemicals affect the function of the immune system; this was documented, for instance, in a volume titled *Biological Markers in Immunotoxicology* (National Research Council, 1992). These chemicals themselves may elicit immune responses, or they may modulate the response of the immune system to different antigens. Such modulation may be manifested by immunosuppression, hypersensitivity to antigens, or autoimmunity. A recent volume on immunotoxicology may be consulted for further details (Dean et al., 1994).

Although not all immune phenomena involve cell membranes, molecules associated with these membranes play major roles in immune reactions. This is referred to in reference to the location of tissue-specific antigens (see Chapter 4) and generally to processes such as cell recognition and communication. The production of immunoglobulin (IgE) elicited by exposure to certain antigens provides a good example of the involvement of cell membranes in immune phenomena. The protein subsequently becomes bound to specific membrane receptors on basophils in the circulation and on mast cells in the tissues. In consequence, these cells become sensitized to the antigen and react with it on further exposure. Surface binding of specific antigens also may transform cells into targets of killer T cells. In general, T lymphocytes recognize antigens expressed on the surface of "antigen-presenting" cells.

Macromolecules and haptens associated with cell membranes thus actively participate in many functions of the immune system. Again, it is therefore not surprising to find that some immunotoxic agents do exert their effect at that site. The extent of membrane involvement in immunotoxic reactions is illustrated subsequently by selected instances. No attempt, however, can be made here to review exhaustively the broad and growing field of immunotoxicology and its relation to membrane function.

An example of immunotoxic action at membranes, already mentioned in section 5-7, is the effect of the immunosuppressive drug cyclosporin on the properties of subcellular membranes. A typical example of immunotoxicity involving the cell membrane is described in the report of Kiremidjian-Schumacher et al. (1981) that heavy metals depress antigen binding. The sensitizing action of nickel on human keratinocytes in allergic contact dermatitis has been attributed to its ability to induce the expression of intercellular adhesion molecule–1 (Gueniche et al., 1994), a surface molecule required for the interaction between antigen-presenting cells and T cells. It is likely that the metals in this case react directly with membranes. The possibility of intracellular mediation in the production of an immunological lesion at the membrane, however, cannot always be excluded (see section 5-3).

Among toxic effects of mercury on humoral immune responses is the depression of surface-membrane immunoglobulins in B lymphocytes of exposed workers (Queiroz and Dantas, 1997). The metal also abolishes K efflux from these cells (Gallagher, Noelle, and McCann, 1995). Lawrence and McCabe (1996) reviewed hypothetical models for the interaction between Pb and the processing and presentation of antigen at the membrane of CD4$^+$T cells. Figure 5-3 shows how the metal might react with the membrane of an antigen-presenting cell: It could directly block a peptide associated with the major histocompatibility complex, or it might directly or indirectly modify the complex. Alternatively, lead might modify an accessory molecule or block the binding of foreign peptides. The original review should be consulted for further details.

The various immunotoxic effects of Cd specifically involving the cell membrane are reviewed by Exon and Koller (1986). This metal interferes with cell-mediated

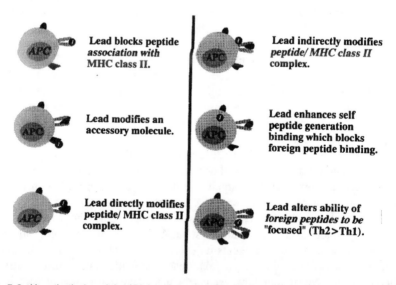

FIG. 5-3. Hypothetical models of Pb interference with cellular handling of antigens. L, metal; APC, antigen-presenting cell; MHC, major histocompatibility. (From Lawrence and McCabe, 1996; with permission.)

immunity, apparently by depressing the reaction of antigen with the cell membrane, a possibility mentioned previously (Kiremidjian-Schumacher et al., 1981). Macrophage function is affected in mice following oral treatment with Cd (Knutson et al., 1980); in particular, the ability of reticuloendothelial cells to bind immune complexes is decreased. Clearance of other macromolecules from blood remained unaffected, however, suggesting that the mechanism of the Cd effect consists of perturbation of specific protein receptors on the cell membrane. Such an action of Cd recalls the previously cited Cd inhibition of the endothelin receptor in human placenta (see section 5-5). Both these effects of Cd are likely to result from direct interactions between the metal and the receptors.

Autoimmune lesions may be elicited if a toxicant modifies tissue constituents in such a manner that they are rejected as foreign by the organism. Some of these modifications presumably involve cell membranes; others result from reactions of readily available molecules in the extracellular compartment. An instance of the latter is the nephrotoxicity of chronic mercury poisoning. This involves autoimmune reactions against altered constituents not of cell membranes but of the glomerular basement membrane, rendered antigenic by the metal. One such potential antigen is the protein laminin, antibodies against which have been detected in the blood of Hg-exposed rats (Chambers and Klein, 1993). Several other heavy metals reacting with tissues also may elicit autoimmune reactions (Bigazzi, 1996); the specific role of the cell membrane in these processes, if any, remains uncertain.

The examples presented here of immune reactions at membranes and the influence of toxicants can do no more than draw attention to an important area in membrane toxicology. More specialized reviews should be consulted for further details.

5-9 CONCLUSIONS

This chapter documents the broad spectrum of effects directly or indirectly exerted by toxic agents on the function of biologic membranes; emphasis is placed on toxic rather than therapeutic or pharmacologic interactions. Both outer plasma membranes and membranes limiting subcellular organelles can serve as targets of toxic action. The membrane functions affected range from specific solute channels to complex, solute movement–related processes such as volume homeostasis, from pinocytosis and membrane traffic to immune phenomena.

Although membrane sites of action at one time were sought for most hormones, drugs, and toxicants, this proved to represent a somewhat simplistic approach to the study of how these agents act; nevertheless, many xenobiotics do affect membrane function. Such effects may result from a direct interaction between the xenobiotics and critical membrane sites; alternatively, the effect may be an indirect consequence of reactions elsewhere in the cell or the body. An instance is the abolition of transmembrane sodium gradients, which secondarily inhibits the coupled transport of solutes like glucose. In practice, it is often difficult to draw clear distinctions between direct and indirect membrane effects.

Toxic actions exerted indirectly on the membrane are mostly relatively nonspecific. An obvious example is the inhibition of energy-requiring processes at the membrane by depression of ATP production in the mitochondria. In contrast, numerous examples have been cited of how toxic agents can react directly at the membrane and thereby induce either specific or more general functional lesions. An example of direct and specific effect is the ouabain inhibition of transmembrane Na transport. A direct but nonspecific membrane effect is that exerted by an agent causing peroxidative or lipolytic changes in the membrane structure. If studied in suitable experimental preparations, membrane toxicology can provide adequate explanations for the action of many, although certainly not all, toxic xenobiotics.

REFERENCES

Abrami, L., V. Berthonaud, P.M.T. Deen, G. Rousselet, F. Tacnet, and P. Ripoche. 1996. Glycerol permeability of mutant aquaporin 1 and other AQP-MIP proteins: Inhibition studies. *Pflugers Arch.* 431:408–414.
Abramson, J.J., E. Buck, G. Salama, J.E. Casida, and I.N. Pessah. 1988. Mechanism of anthraquinone-induced calcium release from skeletal muscle sarcoplasmic reticulum. *J. Biol. Chem.* 263:18750–18758.
Abramson, J.J., and G. Salama. 1989. Critical sulfhydryls regulate calcium release from sarcoplasmic reticulum. *J. Bioenerg. Biomembr.* 21:283–294.
Adams, M.E., and G. Swanson. 1994. Neurotoxins. *Trends Neurosci.* 17 (suppl.):1–27.
Assandri, R., and M. Mazzanti. 1997. Ionic permeability on isolated mouse liver nuclei: Influence of ATP and Ca^{++}. *J. Membr. Biol.* 157:301–309.
Atchison, W.D., and M.F. Hare. 1994. Mechanisms of methylmercury-induced neurotoxicity. *FASEB J* 8:622–629.
Aw, T.Y., B.S. Andersson, and D.P. Jones. 1987. Mitochondrial transmembrane ion distribution during anoxia. *Am. J. Physiol.* 252:C356–C361.
Beavis, A.D. 1992. Properties of the inner membrane anion channel in intact mitochondria. *J. Bioenerg. Biomembr.* 24:77–90.
Bevan, C., and E.C. Foulkes. 1989. Interaction of cadmium with brush border membrane vesicles from the rat small intestine. *Toxicology* 54:297–309.
Bevan, C., E. Kinne-Saffran, E.C. Foulkes, and R.K.H. Kinne. 1989. Cadmium inhibition of L-alanine transport into renal brush border membrane vesicles isolated from the winter flounder. *Toxicol. Appl. Pharmacol.* 101:461–469.
Bigazzi, P.E. 1996. Autoimmunity induced by metals. In *Toxicology of metals.*, ed. L. Chang, 835–852. Boca Raton, FL: CRC Lewis.
Broadway, N.M., and E.D. Saggerson. 1997. Effect of membrane environment on the activity and inhibitability by malonyl-CoA of the carnitine acyltransferase of hepatic microsomal membranes. *Biochem. J.* 322:435–440.
Carrasquer, G., and M. Schwartz. 1996. Effect of acetazolamide and melittin on polarization of the frog gastric mucosa proton pump. *Proc. Soc. Exp. Biol. Med.* 213:258–261.
Catterall, W.A. 1992. Cellular and molecular biology of voltage-gated sodium channels. *Physiol. Rev.* 72(suppl. 4):S15–S48.
Chambers, B.J., and N.W. Klein. 1993. Role of laminin autoantibodies on the embryotoxicity of sera from mercuric chloride treated brown Norway rats. *Reprod. Toxicol.* 7:333–341.
Chavez, E., E. Melendez, C. Zazueta, H. Reyes-Vivas, and S.G. Perales. 1997. Membrane permeability transition as induced by dysfunction of the electron transfer chain. *Biochem. Mol. Biol. Int.* 41:961–968.
Clemens, D.L., C.M. Halgard, J.R. Cole, R.M. Miles, M.F. Sorrell, and D.J. Tuma. 1996. Impairment of asialoglycoprotein receptor by ethanol oxidation. *Biochem. Pharmacol.* 52:1499–1505.
Cohen, M.D., J. Zelikoff, Q. Qu, and R.B. Schlesinger. 1996. Effects of ozone upon macrophage-interferon interaction. *Toxicology* 114:243–252.
Criswell, K.A., and R. Loch-Caruso. 1995. Lindane-induced elimination of gap junctional communication in rat uterine myocytes is mediated by an arachidonic acid-sensitive, cAMP-insensitive mechanism. *Toxicol. Appl. Pharmacol.* 135:127–138.

Cuppoletti, J., and A.J. Abbott. 1990. Interaction of melittin with the Na,K-ATPase: Evidence for a melittin-induced conformational change. *Arch. Biochem. Biophys.* 283:249–257.

Dean, J.H., M.I. Luster, A.E. Munson, and I. Kimber, eds. 1994. *Immunotoxicology and immunopharmacology.* 2d ed. New York: Raven Press.

El Azzouzi, B., G.T. Tsangaris, O. Pellegrini, Y. Manuel, J. Benveniste, and Y. Thomas. 1994. Cadmium induces apoptosis in a human T-cell line. *Toxicology* 88:127–139.

Endo, T., M. Sakata, and Z.A. Shaikh. 1995. Mercury uptake by primary cultures of rat renal cortical epithelial cells: I. Effects of cell density, temperature and metabolic inhibitors. *Toxicol. Appl. Pharmacol.* 132:36–43.

Exon, J.H., and L.D. Koller. 1986. Immunotoxicity of cadmium. In *Handbook of experimental pharmacology,* vol. 80, ed. E.C. Foulkes 339–350. Berlin: Springer Verlag.

Foulkes, E.C. 1956. Cation transport in yeast. *J. Gen. Physiol.* 39:687–704.

Foulkes, E.C. 1972. Cellular localization of amino acid carriers in renal tubules. *Proc. Soc. Exp. Biol. Med.* 139:1032–1033.

Foulkes, E.C. 1991a. Nature of Cd and Hg effects on epithelial amino acid transport in vivo and role of chelators. *Toxicology* 69:177–185.

Foulkes, E.C. 1991b. Further findings on the mechanism of cadmium uptake by intestinal mucosal cells (step 1 of Cd absorption). *Toxicology* 70:261–270.

Foulkes, E.C., and D. Bergman. 1993. Inorganic Hg absorption in mature and immature rat jejunum: Transcellular and intercellular pathways in vivo and in everted sacs. *Toxicol. Appl. Pharmacol.* 120:89–95.

Foulkes, E.C., and S. Blanck. 1991. Cadmium inhibition of basolateral solute fluxes in rabbit renal tubules and the nature of cycloleucine uptake. *Toxicol. Appl. Pharmacol.* 108:150–156.

Foulkes, E.C., and S. Blanck. 1994. 3-O-methylglucose as probe of cytoplasmic volume. *Life Sci.* 54:439–444.

Fussle, R., S. Bhakdi, A. Sziegoleit, J. Tranum-Jensen, T. Kranz, and H. Wellensiek. 1981. On the mechanism of membrane damage by Staph. aureus α-toxin. *J. Cell Biol.* 91:83–94.

Gallagher, J.D., R.J. Noelle, and F.V. McCann. 1995. Mercury suppression of a potassium current in human B lymphocytes. *Cellular Signalling* 7:31–38.

Garst, J.E. 1995. Is carnitine a key toxicological target [abstract]? *Toxicologist* 15:1.

Ghadimiejad, I., and E.D. Saggerson. 1991. A study of properties and abundance of the components of liver carnitinepalmitoyl transferases in mitochondrial inner and outer membranes. *Biochem. J.* 277:611–617.

Girault, L., A. Boudou, and E.J. Dufourc. 1997. Methylmercury interactions with phospholipid membranes as reported by fluorescence, ^{31}P and [199]Hg NMR. *Biochim. Biophys. Acta.* 1325:250–262.

Giridhar, J., A. Rathinaveln, and G.E. Isom. 1992. Interaction of cadmium with atrial natriuretic peptide receptors: Implications for toxicity. *Toxicology* 75:133–143.

Gorlich, D., and I.W. Mattay. 1996. Nucleocytoplasmic transport. *Science* 271:1513–1518.

Grosso, A., and B.C. De Sousa. 1993. Mercury blockage of apical water channels in toad skin (*Bufo Marinus*). *J. Physiol.* 468:741–745.

Gueniche, A., J. Viac, G. Lizard, M. Charveron, and D. Schmitt. 1994. Effect of various metals on intercellular adhesion molecule-1 expression and tumour necrosis factor alpha production by normal human keratinocytes. *Arch. Dermatol. Res.* 286:66–70.

Gunter, T.E., K.G. Gunter, S.S. Scheu, and C.E. Gavin. 1994. Mitochondrial calcium transport: Physiological and pathological relevance. *Am. J. Physiol.* 267:C313–C339.

Guo, T.L., S. Datar, I.M. Shapiro, and B.J. Shenker. 1997. Mercury induced reduction in mitochondrial transmembrane potential and cleavage of poly(ADP-ribose) polymerase in human T cells [abstract]. *Toxicologist* 36:253.

Hart, B.A. 1978. Transport of cadmium by the alveolar macrophage. *J. Reticuloend. Soc.* 24:363–375.

Henry, T.R., and K.B. Wallace. 1995. Differential mechanisms of induction of the mitochondrial permeability transition by quinones of varying chemical reactivities. *Toxicol. Appl. Pharmacol.* 134:195–203.

Huntley, J.S., and A.C. Hall. 1996. Nature of the cation leak induced in erythrocyte membranes by Kanagawa hemolysin of V. parahemolyticus. *Biochim. Biophys. Acta.* 1281:220–226.

Imesch, E., M. Moosmayer, and B.M. Anner. 1992. Mercury weakens membrane anchoring of Na,K-ATPase. *Am. J. Physiol.* 262:F837–F842.

Jehan, Z., and D.B. Motlag. 1995. Metal induced changes in the erythrocyte membrane of rats. *Toxicol. Lett.* 78: 127–133.

Jourd'Heuil, D., P. Vaananen, and J.B. Meddings. 1993. Lipid peroxidation of the brush border membrane: Membrane physical properties and glucose transport. *Am. J. Physiol.* 264:G1009–G1015.

Jungwirth, A., M. Paulmichl, and F. Lang. 1990. Cadmium enhances potassium conductance in cultured renal epithelioid (MDCK) cells. *Kidney Int.* 37:1477–1486.

Kameyama, Y., and M. Gemba. 1991. Cisplatin-induced injury to calcium uptake by mitochondria in glutathione-depleted slices of rat kidney cortex. *Japn. J. Pharmacol.* 55:174–176.

Katsu, T., M. Muroko, T. Morikawa, K. Sanchika, H. Yamanaka, S. Shinoda, and Y. Fujita. 1990. Interaction of wasp venom mastoparan with biomembranes. *Biochim. Biophys. Acta.* 1027:185–190.

Kim, C., W.C. MacKellar, N. Cho, S.R. Byrn, and D.J. Morre. 1997. Impermeant antitumor sulfonylurea conjugates that inhibit plasma membrane NADH oxidase. *Biochim. Biophys. Acta.* 1324:171–181.

Kim, E., S.S.N. Giri, and I.N. Pessah. 1995. Iron (II) is a modulator of ryanodine-sensitive calcium channels in cardiac muscle sarcoplasmic reticulum. *Toxicol. Appl. Pharmacol.* 130:57–66.

Kinter, W.B., and J.B. Pritchard. 1977. Altered permeability of cell membranes. In *Handbook of physiology: Reactions to environmental agents,* ed. D.H.K. Lee, 563–576. Washington: American Physiological Society.

Kiremidjian-Schumacher, L., G. Stotzky, V. Likhite, J. Schwartz, and R.A. Diickstein. 1981. Influence of cadmium, lead and zinc on the ability of sensitized guinea pig lymphocytes to interact with specific antigen and to produce lymphokine. *Environ. Res.* 24:96–105.

Knutson, D.W., D.L. Vredevoe, K.R. Aoki, E.J. Hays, and L. Levy. 1980. Cadmium and the reticuloendothelial system: A specific defect in blood clearance of soluble aggregates of IgG by the liver in mice given cadmium. *Immunology* 40:17–26.

Koschel, K., N.N. Meisner, and P.W. Tas. 1995. Influence of cadmium ions on endothelin-1 binding and calcium signaling in rat glioma C6 cells. *Toxicol. Lett.* 81:189–195.

Latzkovits, L., H.F. Cserr, J.T. Park, C.S. Patlak, K.D. Pettigrew, and A. Rimanoczy. 1993. Effects of arginine vasopressin and atriopeptin on glial cell volume measured as 3-MG space. *Am. J. Physiol.* 264:C603–C608.

Lawrence, D.A., and M.J. McCabe. 1996. Immune modulation by toxic metals. In *Metal toxicology,* eds. R.A. Goyer, C.D. Klaassen, M.P. Waalkes, 305–337. San Diego: Academic Press.

Lozano, R.M., B.C. Yee, and B.B. Buchanan. 1994. Thioredoxin-linked inactivation of venom neurotoxins. *Arch. Biochem. Biophys.* 309:356–362.

MacFarlane, M.G., and A.G. Spencer. 1953. Changes in the water, sodium and potassium content of rat liver mitochondria during metabolism. *Biochem. J.* 54:569–575.

Marin, R., A.J. Rodriguez, and T. Proverbio. 1992. Partial characterization of the inhibitory effect of lipid peroxidation on the ouabain-insensitive Na-ATPase of rat kidney cortex plasma membranes. *J. Bioenerg. Biomembr.* 24:329–335.

Maruyama, Y., H. Shimada, and J. Taniguchi. 1995. Ca activated K channel in the nuclear envelope isolated from single pancreatic acinar cells. *Pflugers Arch.* 430:148–155.

Matsumoto, K., L. Morita, and S. Murota. 1994. Arachidonic acid metabolism by nuclei of a retinoic acid or vitamin D_3-differentiated human leukemia cell line HL-60. *Prostaglandins Leucotrienes. Essential Fatty Acids* 51:51–55.

Menestrina, G., ed., 1994. Pore-forming toxins and antimicrobial polypeptides. *Toxicology.* 87:1–267.

Morselt, A.F., W.M. Frederiks, J.H.C. Peerebom-Steegeman, and H.A. Van Veen. 1987. Mechanism of damage to liver cells after chronic exposure to low doses of cadmium chloride. *Arch. Toxicol.* 11(suppl.):213–215.

Nagahama, M., K. Michiue, and J. Sakurai. 1996. Membrane-damaging action of Clostridium perfringens alpha toxin on phospholipid liposomes. *Biochim. Biophys. Acta.* 1280:120–126.

Nanda, A., J.H. Brumell, T. Nordstrom, L. Kjeldsen., H. Sengelov, N. Borregaard, O.D. Rotstein, and S. Grinstein. 1996. Activation of proton pumping in human neutrophils occurs by exocytosis of vesicles bearing vacuolar-type H^+-ATPases. *J. Biol. Chem.* 271:15963–15970.

Narahashi, T. 1996. Neuronal ion channels as the target sites of insecticides. *Pharmacol. Toxicol.* 79:1–14.

National Research Council. 1992. *Biological markers in immunotoxicology.* Washington: National Academic Press.

Ngkeekwong, F.C., and L.L.Ng. 1997. Two distinct uptake mechanisms for ascorbate and dehydroascorbate in human lymphoblasts and their interaction with glucose. *Biochem. J.* 324:225–230.

Nomiyama, K. and H. Nomiyama. 1994. Cadmium-induced renal dysfunction. *Deutsche Med. Wochenschr.* (*Japn*) 16:189–201.

Oliveira, E.M., J.B. Rocha, and J.J.F. Sarkiss. 1994. In vitro and in vivo effects of $HgCl_2$ on synaptosomal ATP diphosphohydrolase from cerebral cortex of developing rats. *Arch. Int. Physiol. Biochem. Biophys.* 102:97–102.

Ossola, J.O., and M.L. Tomaro. 1995. Heme oxygenase induction by cadmium chloride: Evidence for oxidative stress involvement. *Toxicology* 104:141–147.

Pisoni, R.L., and J.G. Thoene. 1991. The transport systems of mammalian lysosomes. *Biochim. Biophys. Acta.* 1071:351–373.

Preston, R.L., S.J. Janssen, S. Lu, K.L. McQuade, and L. Beal. 1991. Organic and inorganic mercurial inhibition of taurine transport by the coelemocytes of the marine polychaete, Glycera dibranchiata. *Bull. Mt. Desert. Island. Biol. Lab.* 30:72–74.

Prozialeck, W.C., and R.J. Niewenhuis. 1991. Cadmium disrupts intercellular junctions and actin filaments in LLC-PL1 cells. *Toxicol. Appl. Pharmacol.* 107:81–97, 1991.

Queiroz, M.I., and D.C. Dantas. 1997. B Lymphocytes in mercury-exposed workers. *Pharmacol. Toxicol.* 81:130–133.

Rodriguez-Yoldi, M.J., M. Lluch, and F. Ponz. 1987. Action of mercury on sugar transport across rat small intestine, in vivo. *Revista Espanola de Physiol.* 43:239–244.

Rodriguez-Yoldi, M.J., and F. Ponz. 1987. Inhibition of sugar active transport across rat intestine in vivo by cadmium. *Revista Espanola de Physiol.* 43:39–44.

Rogers, J.R., Y. Qu, T.N. Tanada, T. Scheuer, and W.A. Catterall. 1996. Molecular determinants of high affinity binding of α-scorpion toxin and sea anemone toxin in the S3–S4 extracellular loop in domain IV of the Na+ channel α subunit. *J. Biol. Chem.* 271:15950–15962.

Romero, G., B. Lasheras, S. Suberviola, and E. Cenarruzabeita. 1994. Protective effects of Ca channel blockers in carbon tetrachloride-induced liver toxicity. *Life Sci.* 55:981–990.

Rossi, A., R. Poverini, G. Di Lullo, A. Modesti, A. Modica, and M.L. Scarino. 1996. Heavy metal toxicity following apical and basolateral exposure in the human cell line Caco-2. *Toxicol. in Vitro* 10:27–36.

Rothstein, A. 1959. Cell membrane as site of action of heavy metals. *Fed. Proc.* 18:1026–1035.

Schneeberger, E.E., and R.D. Lynch. 1992. Structure, function and regulation of cellular tight junctions. *Am. J. Physiol.* 262:L647–L661.

Schwartz, Z., P.A. Gates, and E. Nasalzky. 1996. Effect of 17β-estradiol on chondrocyte membrane fluidity and phospholipid metabolism is membrane-specific, sex-specific, and cell maturation–dependent. *Biochim. Biophys. Acta.* 1282:1–10.

Scorrano, L., V. Petronilli, and P. Bernardi. 1997. On the voltage dependence of the mitochondrial permeability transition pore. *J. Biol. Chem.* 272:12295–12299.

Sellinger, M., N. Ballatori, and J.L. Boyer. 1991. Mechanisms of mercurial inhibitors of sodium-coupled alanine uptake in liver plasma membrane vesicles from Raja erinacea. *Toxicol. Appl. Pharmacol.* 1107:369–376.

Shopicky, H.A., D. Zikos, D., E.J. Sukowski, K.A. Fisher, and D.R. Peterson. 1996. Gentamycin inhibits carrier-mediated transport in the kidney. *Am. J. Physiol.* 270:F531–F538.

Silverman, M., M.A. Aganon, and F.P. Chinard. 1970. D-glucose interaction with renal tubule cell surfaces. *Am. J. Physiol.* 218:735–742.

Simons, J.W., S.J. Noga, P.M. Colombani, W.E. Besehorner, D.S. Coffey, and A.D. Hess. 1986. Cyclosporin A, an in vitro calmodulin antagonist, induces nuclear lobulations in human T cell lymphocytes and monocytes. *J. Cell Biol.* 102:145–150.

Solem, L.E., and K.G. Wallace. 1993. Selective activation of the sodium-independent cyclosporin A sensitive calcium pore of cardiac mitochondria by doxorubicin. *Toxicol. Appl. Pharmacol.* 121:50–57.

Strzelecki, T., S. Kumar, R. Khauli, and M. Menon. 1988. Improvement by cyclosporin of membrane-mediated functions in kidney mitochondria. *Kidney Int.* 34:234–240.

Suwalsky, M., M. Benites, F. Villena, F. Aguilar, and C.P. Sotomayor. 1997. The organochlorine pesticide heptachlor disrupts the structure of model and cell membranes. *Biochim. Biophys. Acta* 1326:115–123.

Suwalsky, M., I. Sanchez, M. Bagnara, and C. Sotomayor. 1994. Interaction of antiarrythmic drugs with model membranes. *Biochim. Biophys. Acta* 1195:189–196.

Tedeschi, H., and D.L. Harris. 1955. The osmotic behavior and permeability to non-electrolytes of mitochondria. *Arch. Biochem.* 58:52–67.

Udden, M.H., and C.S. Patton. 1994. Hemolysis and deformability of erythrocytes exposed to butoxyacetic acid, a metabolite of 2-butoxyethanol: Sensitivity in rats and resistance in normal humans. *J. Appl. Toxicol.* 14:91–96.

Uki, M., and T. Narahashi. 1996. Modulation of serotonin-induced currents by metals in mouse neuroblastoma cells. *Arch. Toxicol.* 70:652–660.

Ullrich, K.J., and F. Papavassiliou. 1985. Contraluminal transport of hexoses in the proximal convolution of the rat kidney in situ. *Pflugers Arch.* 404:150–156.

Vansteveninck, J., R.I. Weed, and A. Rothstein. 1965. Localization of erythrocyte membrane sulfhydryl groups essential for glucose transport. *J. Gen. Physiol.* 48:617–632.

Verbost, P.M., G. Flik, R.A. Lock, and S.E. Wendelaar-Bonga. 1988. Cadmium inhibits plasma membrane calcium transport. *J. Membr. Biol.* 102:97–104.

Verity, M.A. 1996. Nervous system. In *Metal toxicology*, eds. R.A. Goyer, C.D. Klaassen, M.P. Waalkes, 199–235. San Diego: Academic Press.

Viarengo, A. 1994. Heavy metal cytotoxicity in marine organisms: Effects on calcium homeostasis and possible alteration of signal transduction pathways. *Adv. Comp. Environ. Physiol.* 20:85–110.

Vig, P.J.S., S.N. Pentyala, C.S. Chetty, B. Rajanna, and D. Desaiah. 1994. Lead alters inositol polyphosphate receptor activities: Protection by ATP. *Pharmacol. Toxicol.* 75:17–22.

Wada, K., Y. Fujii, H. Watanabe, M. Satoh, and Y. Furuichi. 1991. Cadmium directly acts on endothelin receptor and inhibits endothelin-binding activity. *FEBS Lett.* 285:71–74.

Wallin, M., and B. Hartley-Asp. 1993. Effects of potential aneuploidy inducing agents on microtubule assembly in vitro. *Mutat. Res.* 287:17–22.

Whiting, P.H. 1993. Cyclosporin A–induced nephrotoxicity. In *Renal disposition and nephrotoxicity of xenobiotics*, eds. M.W. Anders, W. Dekant, D. Henschler, H. Oberleithner, and S. Silbernagl, 235–247. New York: Academic Press.

Williams, P.D., and G.H. Hottendorf. 1985. Effect of cisplatin on organic ion transport in membrane vesicles from rat kidney cortex. *Cancer Treat. Rep.* 69:875–880.

Yu, S.P., C.-H. Yeh, S.L. Sensi, B.J. Gwag, L.M.T. Canzionero, Z.S. Farhangrazi, H.S. Ying, M. Tian, L.L. Dugan, and D.W. Choi. 1997. Mediation of neuronal apoptosis by enhancement of outward potassium current. *Science* 278:114–117.

Yuan, C., and D. Acosta. 1996. Cocaine-induced mitochondrial dysfunction in primary cultures of rat cardiomyocytes. *Toxicology* 112:1–10.

Zara, V., G. Damiano, M.R. Franco, L. Siculella, and G.V. Gnoni. 1996. Partial purification and reconstitution of the tricarboxylate carrier from eel liver mitochondria. *Biochem. Mol. Biol. Int.* 39:369–375.

Zengchua, M., J.B. Meddings, and S.L. Lee. 1994. Membrane physical properties determine cardiac β-adrenergic receptor function in cirrhotic rats. *Am. J. Physiol.* 267:G87–G93.

Zucker, R.M., K.H. Elstein, R.E. Eastering, and E.J. Massaro. 1988. Metal-induced alteration of the cell membrane/cytoplasm complex studied by flow cytometry and detergent lysis. *Toxicology* 53:69–78.

6

Membrane Transport of Toxicants

6-1 INTRODUCTION

Reference has been made repeatedly in earlier chapters to the determining role played by cell membranes in the handling of nonlipophilic xenobiotics in the body. The role is critical for such toxicants because they generally do not passively permeate through cell membranes. In consequence, their absorption into the body, their distribution to different organs, and their accumulation at intracellular target sites, as well as frequently their extrusion from cells, in general require mediation by more-or-less specific facilitated diffusion or active transport mechanisms in cell membranes. Finally, excretion of xenobiotics from the body, for instance through the liver or the kidneys, also is controlled by cell membranes: A polar solute is not reabsorbed from the glomerular filtrate or secreted into the renal tubule in the absence of special membrane transport mechanisms. In other words, membrane properties and function largely determine the toxicokinetics of these toxic agents. In contrast, uptake, distribution, and excretion of lipophilic solutes, to which membranes generally are relatively freely and passively permeable, are little affected by membranes.

Although function of cell membranes can be directly or indirectly altered by many toxic agents (see Chapter 5), the transient membrane contact made by toxic xenobiotics during their transmembrane movement, as described above, by itself need not be associated with lasting effects on membrane function. Nevertheless, because of the important role of membranes in toxicokinetics, membrane transport processes for toxic xenobiotics must be included in a discussion of the role of biologic membranes in toxicology. Where membranes mediate cellular uptake of toxic agents, the normal protective role of the membrane (see Chapter 2) is replaced by its opposite, a permissive role, as further discussed below. This chapter reviews the role of cell membranes in determining the toxicokinetic behavior of xenobiotics, with special emphasis on their uptake into different cell types.

A number of instances are described in which the access of toxic agents to their intracellular target is mediated by mechanisms at the cell membrane. The toxicity of several nephrotoxicants, for instance, is associated with their active accumulation by tubule cells (see section 6-6). An important corollary of such processes is that inhibition of the uptake system should protect the cell against intoxication. Cells also can protect themselves, as briefly mentioned, against the action of certain toxic agents and waste products by catalyzing their active extrusion; the carrier systems involved may be constitutive, or they may be activated on exposure. Inhibition of such export carriers renders a cell more sensitive to their substrates. This is a field with major therapeutic implications and is analyzed further in Chapter 7.

6-2 GENERAL ROLE OF CELL MEMBRANES IN TOXICOKINETICS

The field of toxicokinetics encompasses the study of how the effects of a toxic agent are determined by the rate and route of its uptake, its distribution in the body, its retention in various tissues, its metabolic activation or detoxification, its sequestration as in the case of bone-seeking metals, and finally, the rate of its excretion. As pointed out, the toxicokinetics of a lipid-insoluble xenobiotic, therefore, are influenced greatly by its ability to cross cell membranes. Such a determining role is diminished to the extent that a solute can passively cross cellular barriers such as, for example, the placenta, the intestinal mucosa, or renal tubular epithelium. Such passive permeation may reflect the presence of leaky junctions between cells or of polar pores in the cell membrane.

Even if solutes cross cell barriers along transcellular pathways, however, there is little resistance to the movement of lipid-soluble compounds, capable of nonionic passive diffusion into and out of cells. For instance, compounds such as cyanide or ammonia, both of which are partly undissociated near physiologic pH values, or vapors and gases like elemental mercury ($Hg°$), H_2S or CS_2, all can move across cell membranes passively and relatively freely. These facts are illustrated dramatically by the high central nervous system toxicity of dimethylmercury as compared with that of Hg^{++}.

Measurement of the intrinsic permeability of a complex membrane, defined as unidirectional solute flux per unit area, requires knowledge of this area. This may be difficult to obtain, especially for highly folded or ciliated membranes. Yet it is only on the basis of such measurements that one can compare, for instance, the intrinsic solute permeability of the mature and the neonatal gut. Although an estimate of absorbing area in the intestine can be obtained morphometrically on fixed tissue, a functional technique for its estimation in vivo also has been developed. This procedure takes advantage of the simultaneously measured passive flux of a second solute, which must be both water- and lipid-soluble (Foulkes, Mort, and Buncher, 1991). Such a solute is ethanol; at a given temperature the exponential rate constant of its passive uptake by a tissue depends only on the size of the absorbing area, not on the composition of the membrane or other factors. Doubling of the contribution of aqueous pores to the total absorbing area, for instance, has little effect on the transmembrane flux of ethanol. In other words, for a given concentration gradient at a constant temperature, the rate of ethanol uptake is proportional only to the total surface area involved.

It follows from these facts that normalization of solute fluxes across membranes on the basis of simultaneous ethanol flux should yield, in empirical units, a measure of flux per unit membrane area, or of the intrinsic permeability of the membrane. This technique permits the demonstration, for instance, that the handling of intestine during the preparation of everted sacs greatly increases the frequency of aqueous pores; this can be measured by the increased ratio of urea to ethanol fluxes (see Chapter 3). It also becomes possible quantitatively to estimate the influence of such factors as age, diet, and changes in the environment on the intrinsic metal permeability of the intestinal barrier (see section 7-4).

To the extent that movement of solutes from intestinal lumen or renal tubular urine follows primarily a transcellular pathway rather than involving diffusion across more-or-less tight junctions between cells (Foulkes and McMullen, 1987), the first step in absorption must consist of a transient interaction of these solutes with apical cell membranes. Similarly, renal secretion of, for instance, organic anions like PAH across the tubular epithelium into urine is initiated by interaction of the secreted solute with the anion carrier in basolateral cell membranes (Foulkes and Miller, 1959); this is also the site of action of a number of specific inhibitors of PAH transport.

The transient contact between membrane constituents and a toxicant crossing the membrane need not interfere with membrane function. Thus ricin and Shiga toxin, two extremely cytotoxic proteins, exert their effects entirely at intracellular targets (see review by Sandvig and van Deurs, 1996); they reach these targets by endocytosis following their binding to the membranes. In other instances, in contrast, reaction of a toxicant with the membrane leads to immediate functional changes, such as are discussed for Hg^{++} in Chapter 5.

Transient electrostatic interaction between the cell membrane and many polyvalent cations, such as those of Cd and other heavy metals, at least transiently alters charge distribution on the membrane. Possible effects of such alterations are illustrated by the observation that removal of external negatively charged sialic acid residues from the membrane of Chinese hamster ovary cells alters the electric field sensed by channel gating elements (Bennett et al., 1997); a large fraction of the fixed anionic charges on cells, as mentioned in Chapter 4, is sensitive to hydrolysis by neuraminidase. Similarly, changes in charge distribution on a cell membrane caused by electrostatic binding of polyvalent cations may affect transmembrane potentials and electrochemical activity gradients and thereby alter gating of voltage-sensitive channels and ion fluxes down the gradient.

The significance of such changes in membrane function is reduced, however, by the observation that the rapid binding of Cd^{++} to the brush border of the jejunal mucosa does not lead to overt and immediate effects on membrane function. Thus, the metal inhibits transport of the amino acid analogue cycloleucine at the intestinal brush border only after a pronounced initiation delay (see section 5-3). Similarly, in purified membrane vesicles exposed even to a relatively high concentration of $CdCl_2$ (0.1 mM), maximal inhibition of the initial velocity of glucose uptake was seen only after about 10 minutes (Bevan and Foulkes, 1989). Several possible explanations are offered in Chapter 5 for the initiation delay of Cd, including difficult access to its target site or the initiation of a slow and progressive reaction such as lipid peroxidation that nonspecifically alters structure and function of the cell membrane.

The fact that Cd and some other toxic metals thus react with membrane constituents during cellular uptake provides no proof for the existence of specific membrane transport systems mediating the absorption of toxic and nonessential heavy metals. Thus, the apparent saturability of cadmium uptake by enteric cells results from nonspecific charge neutralization on the membrane (see section 6-3). The temperature dependence of metal uptake also at times has been cited inappropriately as evidence for the contribution of an energy-requiring process (Endo, Sakata, and Shaikh, 1995; Hart, 1978). Actually, the observed Q_{10} values can be calculated to approximate 1.4, significantly lower than what would be expected from most facilitated processes (see section 5-2); it must be remembered that even physical diffusion in an aqueous solution exhibits a Q_{10} near 1.2. The conclusion, therefore, that a Q_{10} of approximately 1.2 for Cd uptake by a human liver cell line indicates carrier mediation, as reported by Souza et al. (1996), is not justified.

A Q_{10} value of about 1.4 also was observed for the internalization of Cd bound to the brush border membrane of rat enterocytes (Foulkes, 1991a) and was attributed possibly to the expected change in membrane fluidity with temperature. The preliminary binding of Cd to the membrane is characterized by a Q_{10} of less than 1.1, and that of Hg by a value of 1.2 (Foulkes and Bergman, 1993). As is seen in section 6-3, temperature dependence and apparent saturation kinetics of cellular Cd uptake, at least in enteric cells, can be accounted for readily without having to invoke specific Cd transporters.

In any case, as previously emphasized (Foulkes, 1985), development of mechanisms designed specifically to catalyze absorption of toxic and nonessential heavy metals would possess no selective survival value and therefore would be biologically meaningless. Of course, it can be argued that the specificity of transport systems designed for essential metals such as Zn or Ca are sufficiently broad to permit mediation of some uptake of toxic and nonessential metals. No evidence for competition between Cd and Zn or Ca was observed, however, during studies on metal absorption in the rat jejunum in the author's laboratory (Foulkes, 1985); the significance of this negative finding is further developed in the next section. A different conclusion was drawn for Cd and Zn interaction, however, by Hoadley and Cousins (1985), studying the whole length of the gastrointestinal tract rather than only the jejunum. Competition between Cd and Ca also has been suggested to occur in other cell types (e.g., Hinkle, Kinsella, and Osterhoudt, 1987; see section 6-8).

More compelling evidence than that for carrier-mediated transport of inorganic heavy metals across cell barriers has been found for some metal–organic compounds and other toxicants. Examples are the renal secretion of mercurial diuretics

(Borghgraef, Kessler, and Pitts, 1956) and cisplatin compounds (Daley-Yates and McBrien, 1985), the transport of methylmercury across the blood–brain barrier (see subsequent text), the basolateral uptake of paraquat by renal tubules (see section 6-6), and the accumulation of cephaloridin at its intracellular site of action in the kidney (section 6-5). The carrier systems responsible for these processes in some instances have been identified with those responsible for transport of physiologic solutes such as amino acids or organic anions. Finally, toxic agents may be actively extruded across the cell membranes by broadly specific and well-characterized transporters. Such active export of toxic agents from cells can render them resistant to intoxication; this represents a special case of toxicant transport and is further considered in Chapter 7.

Impermeability of membranes, as pointed out, restricts the distribution of toxicants in the body. Inorganic mercury, for instance, does not readily cross the blood–brain barrier; in contrast, the central nervous system is a target for the significantly more lipid soluble mercury vapor or for monomethyl mercury (MMHg) (see section 6-7). It must be added, however, that MMHg is still a charged molecule and that the mercurial is likely to be circulating in plasma ligated to compounds like cysteine rather than in the free form. The cysteine complex has been shown to serve as substrate of the transporter for neutral L-amino acids. Lipid solubility per se, therefore, cannot provide a full explanation for the ready uptake of MMHg across the blood–brain barrier or in the intestine. Dimethyl mercury, finally, is a volatile and very lipid-soluble compound that readily crosses cell membranes. Its extremely high central nervous system toxicity is related not only to its rapid transfer across the blood–brain barrier but also to its subsequent demethylation to MMHg; as such it binds essentially irreversibly to various brain constituents.

Membrane transport of toxicants is reviewed in this chapter under the headings of specific tissues or cell types. Although different cells may transport a certain toxicant by different mechanisms, there is also evidence for significant conservation among such mechanisms. A good example are the P-ATPases catalyzing ion movement in many tissues (Solioz and Vulpe, 1996). They are large monomeric membrane molecules inhibited by micromolar concentrations of vanadate and forming acylphosphate intermediates. Among them is a subfamily, the so-called CPx type, whose members catalyze transmembrane movement of heavy metals and are distinguished by special motifs. Although the Cd and Hg pumps have been found mostly in microorganisms, P-ATPases transporting Cu have been described in mammals.

All these results indicate the extent to which cell membranes, in addition to potentially serving as targets of toxic action, also are involved in the movement of toxicants in the body.

6-3 INTESTINAL TRANSPORT OF TOXICANTS

Although transdermal absorption of lipid-soluble toxicants can be very significant, outside the occupational environment nonvolatile compounds generally are taken into the body via the gastrointestinal tract. A variety of membrane mechanisms contribute to the intestinal absorption of these substances. In addition, transepithelial secretion into the intestinal lumen has also has been repeatedly observed for certain toxicants, including heavy metals. Because a detailed discussion of intestinal solute transport lies beyond the scope of this section, reference for further details should be made to general volumes such as the collection of reviews edited by Csaky (1984).

Many specific membrane carrier mechanisms have been identified in the intestinal mucosa. They mediate uptake of required food constituents including sugars, amino acids, fats, vitamins, inorganic electrolytes, and other solutes. The specificity and physiologic significance of the various absorption mechanisms involved is indicated by the observation that they generally are under homeostatic control; V_{max} is increased in deficiency states. In contrast, the concept of homeostasis can hardly be expected to apply to the uptake of nonessential and toxic solutes like some of the heavy metals. Uptake of Cd, for instance, is not influenced by its body burden (Cotzias, Borg, and Selleck, 1961).

No special mediation, of course, is required for lipid-soluble compounds, or for any substance that might be able to cross the intestinal barrier by passive diffusion. Such passive movement may occur both through the cells or between the cells and is not likely to be structure-specific. Even entirely passive processes, however, as pointed out previously, may be expected to vary significantly with temperature. This fact may reflect the temperature-dependence of physical diffusion, of membrane fluidity, and perhaps of other factors. As emphasized in the previous section, a relatively small temperature-dependence must not be attributed automatically to a requirement for metabolic energy or for carrier action.

To the extent that specific mechanisms are involved in the transmural movement of solutes out of the intestinal lumen, as in the case of glucose and amino acids, the process may be assumed to follow a transcellular rather than an intercellular pathway. Although there is little evidence to support function of specific carriers for the toxic heavy metals, results obtained with rat jejunum also point to a transcellular route of absorption. Thus, in the case of Cd, the process is depressed by the presence of endogenous metallothionein in the cells (Foulkes and McMullen, 1986a); the time required for movement of the metals from lumen into the portal vein is a direct function of their affinity for metallothionein (Foulkes and McMullen, 1987). Further, the kinetics of the process indicate that the tissue pool of Cd through which the metal passes is coextensive with total tissue Cd (Foulkes and McMullen, 1987); the metal, of course, is not likely to accumulate in extracellular pools in the tissue. The apparent transcellular route of metals like Cd across the intestinal barrier in the jejunum implies that, as a first step in their absorption, metals must interact with the apical cell membranes (see section 6-2).

Two possible mechanisms have been suggested to explain the tranfer of nonessential heavy metals across mucosal cell membranes. The first hypothesis, as in section 6-2, has the metal reacting with relatively nonspecific transport systems primarily dedicated to the uptake of essential nutrients or specific metabolites, as would be the case if Cd were transported by Zn or Ca carriers. This hypothesis, however, necessitates a competitive interaction between Cd and Zn or Cd and Ca; these metals do interact in perfused segments of jejunum in vivo but do not do so in a competitive manner, as further described subsequently. Transport by a regular carrier system also would be expected to exhibit a higher Q_{10} than the value of 1.4 actually observed (see section 6-2). Further, if Cd were absorbed by a Zn carrier system, changes in the homeostatically controlled rate of Zn uptake, which is raised in Zn-deficient rats, should be accompanied by parallel variations in Cd uptake; this was not observed by Foulkes and Voner (1981) in the rat jejunum.

Nevertheless, some suggestive evidence for the existence of heavy metal carriers has been reported. A putative transporter for divalent metals recently was identified in the intestine and other tissues of the rat (Gunshin et al., 1997). It is a protein consisting of 561 amino acid residues, with 12 suggested membrane-spanning domains, and may be involved primarily in active, proton-coupled absorption of Fe^{2+} from the intestine. The transporter also has been shown to react with other divalent metals, including Zn, Cd, Cu, Ni, and Pb. Valberg, Sorbie, and Hamilton (1976) had observed previously an interaction between the gastrointestinal absorption of Cd and Fe.

The second detailed possibility that has been proposed to account for the intestinal uptake of nonessential and toxic heavy metals explains the process without having to make the biologically meaningless assumption of a role for specific metal carriers. Details of the suggested mechanism are considered in other sections but may be summarized as follows. The first step in this scheme is the electrostatic reaction of Cd and other polyvalent cationic metals with binding sites on the outer membrane. Neutralization of the negative membrane charges by all polyvalent cations tested, including polylysine or Cd itself, predictably reduces this binding reaction. The apparent "saturating" effect of high Cd concentrations is fully explained on the basis of charge neutralization and does not require the assumption of a saturable carrier.

The electrostatic binding of the metal to the outer surface of the membrane is temperature-independent and is labelled step 1A of cellular metal uptake (in Fig. 3-2). Portions of the membrane-bound metal subsequently become internalized by step 1B of absorption; the proposed explanation for the small temperature-dependence of this step ($Q_{10} = 1.4$), as mentioned previously, is believed to reflect changes in membrane fluidity. It is interesting to speculate that a cation diffusion facilitator protein, as reviewed by Paulsen and Saier (1997), may be involved in binding and internalization of heavy metals in the intestine.

The mechanism of the intestinal uptake of toxic cationic heavy metals, as outlined, has been previously reviewed (see, e.g., Foulkes, 1994) and possesses the advantage of being able to explain the apparent saturability of metal uptake in rat jejunum, its inhibition by polyvalent cations, and its small temperature-dependence without having to make the biologically meaningless postulate that specific carriers have evolved to mediate uptake of toxic and nonessential heavy metals. It must be emphasized, however, that conclusions drawn on the nature of solute transport in the rat jejunum are not necessarily applicable to other cell types (see section 6-8).

Both intestinal loops and everted sacs have been applied extensively to the study of metal absorption. Use of sacs, however, suffers from the double disadvantage, as discussed in section 3-5, that the submucosal tissue avidly retains metals and that apparent artifacts of preparation greatly increase the contribution of aqueous pores to the total absorbing area in sacs. In fact, the increased frequency of aqueous pores in part may be equivalent to greater availability of intercellular diffusion pathways, leading to a correspondingly smaller significance of transcellular pathways. In that case, the linear model shown in Fig. 3-2 does not describe adequately intestinal solute absorption; this is more appropriately represented as the parallel kinetic model. Implicit application of the wrong model has led to some unjustified interpretations, such as the claim that EDTA decreases Cd uptake at the apical side but increases its extrusion on the basolateral side (Kojima and Kiyozumi, 1974). The series model cannot readily account for increased appearance of Cd in the serosal fluid, while apical uptake into the cells is inhibited.

Using the appropriate linear model from Fig. 3-2 to describe the uptake of heavy metals from the intestinal lumen, their transcellular diffusion pathway, and their basolateral extrusion, the absorption process can be separated into steps 1A and 1B at the apical cell membrane, and the basolateral extrusion step 2. The first reaction of cationic species of heavy metals with the brush border (step 1A), as stated previously, appears to consist of their electrostatic binding to anionic charges on the membrane, followed by internalization of a portion of the bound metal (step 1B of cellular uptake). The non-competitive interaction between metals at the level of step 1 is illustrated in Fig. 6-1.

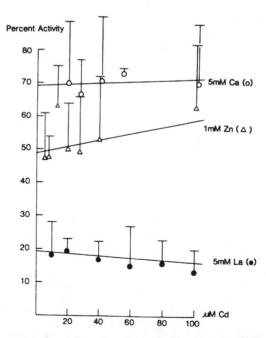

FIG. 6-1. Noncompetitive interaction of Cd with other heavy metals in rat jejunum. (From Foulkes, 1985; with kind permission from Elsevier Science Ireland Ltd. Bay 15K, Shannon Industrial Estate, Co. Clare, Ireland.)

Note that fractional inhibition of Cd uptake by selected cations was independent of the concentration of Cd over a 10-fold concentration range. Such noncompetitive inhibitory action of a heavy metal also is seen in Fig. 5-3 relating to binding of endothelin to its membrane receptor. Interaction between Cd and Zn in the intestine not involving competition for a common carrier is suggested further by the observation that increased Zn absorption during dietary Zn deficiency is not accompanied by changes in jejunal Cd absorption (Foulkes and Voner, 1981). A contrary conclusion was drawn by Hoadley and Cousins (1985) who reported competition between the two metals for intestinal absorption. Reasons for this discrepancy are not clear but may be related to use of different preparations: Isolated segments of jejunum were studied by Foulkes and Voner; Hoadley and Cousins used essentially the whole length of the intestinal tract.

An electrostatic (charge-neutralizing) rather than a saturation effect of excess Cd during Cd transport in the perfused rat jejunum also is supported by the finding that uptake of ^{109}Cd from 5 μM $CdCl_2$ was reduced to 28% $\pm$ 12% of control by the addition of 250 μM $CdCl_2$, and to 29% $\pm$ 16% by 250 μM $ZnCl_2$ (Foulkes, 1991a); in other words, excess Cd and Zn exerted the same effect. Similar findings previously had been reported on the apparent saturation of Cd uptake with either Cd or Ni (Foulkes and McMullen, 1987). An alternative but less likely explanation based on true saturability of metal transporters would have to postulate a system with equal affinity for the three different metals (Cd, Zn, and Ni).

An a priori argument can be made further against equating Ca and Zn inhibition of Cd tranport with involvement of Ca and Zn carrier systems in Cd absorption: By analogy, it would be necessary to attribute the inhibitory action of lanthanum, lead, polylysine, and other cations to competition for their specific carrier systems. Existence of a specific system for transport of, for example, La^{3+} or polylysine, has never been suggested and indeed seems unlikely. Instead, the nonspecificity of the interaction between Cd and all other polyvalent cations tested is compatible with the hypothesis that these inhibitory cations act not by competing for a common carrier but by nonspecifically neutralizing the anionic membrane charges required for Cd binding. The nature of the anionic charges is considered in Chapter 4; sialic acid residues constitute a major fraction of the total number of anionic sites on the membrane.

The general inhibition of Cd absorption by polyvalent cations, attributed to neutralization of fixed anionic charges required for binding of Cd, must be contrasted with the opposite effect of cations on Hg absorption (Foulkes and Bergman, 1993), which presumably reflects the presence of Hg as a polychloride anion. Trivalent La, for instance, stimulates uptake of Hg and inhibits that of Cd; apparently, neutralization of fixed anionic charges depresses the binding of cationic Cd but facilitates the approach of anionic Hg to its membrane binding sites. The anion channel inhibitor DIDS, however, does not interfere with Hg uptake by enteric cells. Nor is there evidence in these cells of the movement of other metals through anion channels, as described, for instance, for erythrocytes. It is of interest in this connection that a Cd-related decrease in negative membrane charges has been invoked to explain the loss of electrophoretic mobility

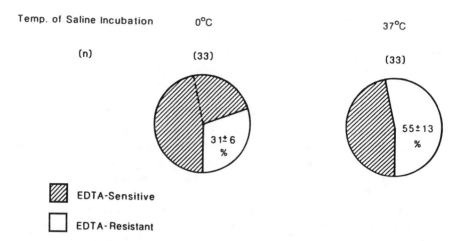

Temp. of Saline Incubation   0°C   37°C

(n)   (33)   (33)

31±6 %

55±13 %

▨ EDTA-Sensitive

☐ EDTA-Resistant

FIG. 6-2. Functional compartmentation of membrane-bound Cd. (From Foulkes, 1988; with kind permission from Elsevier Science Ireland Ltd., Bay 15K, Shannon Industrial Estate, Co. Clare, Ireland.)

of macrophages exposed to the metal in vitro (Nelson, Kiremidjian-Schumacher, and Stotzky, 1982).

Multiple classes of chemical residues on the cell membrane potentially can bind heavy metals. They include, of course, the anionic groups apparently involved in step 1A of intestinal Cd absorption. Besides these fixed anionic charges, other membrane groups with significant affinity for metals include sulfhydryl and imidazole residues. The presence of different metal binding sites on the membrane is compatible with the observation that only a portion of the bound metal subsequently participates in step 1B, the internalization of bound metal (Fig. 6-2).

The approach to this problem illustrated in Fig. 6-2 utilizes the finding that surface-bound metals can be removed readily by short exposure to low concentrations of a suitable chelator, which itself does not readily enter cells; such a compound, for instance, is EDTA for Cd (Foulkes and McMullen, 1987) or dimercaptosuccinate for Hg (Foulkes and Bergman, 1993).

It is possible with short exposure to cold 1 mM EDTA, for instance, to separate the Cd on the surface from that which has been transferred more deeply into the membrane or into the cell interior. In the experiments shown in Fig. 6-2, 69% of Cd taken up by everted sacs in 10 seconds remained on the surface, even after further incubation in Cd-free medium at 0°C. The remaining 31% of tissue Cd was not accessible to EDTA unless the membrane had been damaged first by detergents or by freezing and thawing. The normally EDTA-resistant compartment therefore operationally is defined as the internalized pool. If the second incubation, in Cd-free saline, took place at 37°C instead of 0°C, the internalized Cd pool increased from 31% to 55% of tissue Cd, with only 45% remaining on the surface; prolonging the incubation did not further reduce surface Cd. By definition, the internalization of (69%−45%)/69%, or about one-third of the membrane Cd, represents step 1B of Cd uptake.

Step 1B, as functionally defined here, is temperature-sensitive. The observed Q_{10} value of the process equals only 1.4 and cannot prove participation of an energy-requiring process. The contrary conclusion has been drawn in the past repeatedly but inappropriately in various systems (see section 6-2). Instead, the effect of temperature may reflect primarily changes in membrane fluidity. An important role of this property in metal transport is discussed in section 4-3; for instance, it strongly influences passage of xenobiotics across the buccal mucosa (Aungst, 1996). In summary, there is little evidence to support the hypothesis that active transport is involved in the uptake of heavy metals by enteric and other cells. In any case, given the electrically negative interior of cells, and the high concentration of metal ligands, it is very unlikely that uptake of cationic heavy metals into cells ever proceeds against an electrochemical activity gradient and therefore that it would ever require an involvement of active transport.

The apparent functional compartmentation of membrane-bound Cd illustrated in Fig. 6-2, as stated previously, may reflect binding of the metal to different membrane components. It must be pointed out, however, that no kinetic evidence for heterogeneity of binding sites was obtained in studies on equilibration rates of membrane-bound Cd with extracellular Cd (Foulkes, 1991). The EDTA-resistant Cd need not be located within the cell. This follows from the further observation that, although none of the Cd taken up by purified membrane vesicles is free in solution and therefore must be bound to the membrane (Bevan and Foulkes, 1989), only a portion can be removed readily with EDTA. The inability of portion of the membrane-bound Cd to react with EDTA therefore suggests that the binding site lies relatively deeply buried in the membrane structure, or even on its intravesicular side. If these sites are involved in, for example, solute transport across the membrane, their existence may help explain the finding of an initiation delay before Cd begins to inhibit membrane function in brush border membrane vesicles (see, e.g., Bevan et al., 1989) as well as in intact tissue.

In erythrocytes, anionic metal complexes like polychlorides of Hg^{++} enter the cells by a mechanism sensitive to the anion-channel inhibitor diisocyanatostilbene disulfonate (DIDS) (Simons, 1986; Lou, Garay, and Alda, 1991). The binding of Hg^{++} to jejunal brushborder also involves an anionic complex, as discussed previously. Nevertheless, movement of the metal into enteric cells is not inhibited by DIDS (Foulkes and Bergman, 1993). Instead of relying on anion channels, uptake of Hg in these cells probably involves first a reaction with the cell membrane, although clearly not with the anionic sites responsible for Cd binding. The further finding that bound Hg cannot be dissociated with EDTA suggests that the reaction involves sites like sulfhydryl groups possessing a higher affinity for the metal than does EDTA. Hg binding can be reversed, however, by short exposure to extracellular dimercaptosuccinate (Foulkes and Bergman, 1993); this provides a tool for separating surface-bound from internalized Hg.

Cadmium has been reported competitively to inhibit ATP-requiring Ca transport by basolateral membrane vesicles of rat duodenum (Verbost et al., 1988). No evidence for any carrier participation in step 2 of Cd absorption, however, was found by Foulkes (1993). In any case, at the apical cell membranes in jejunum the apparent absence of competition between Cd and other metals including Ca (see Fig. 6-1) diminishes the likelihood that Cd in this system is absorbed by specific Ca carriers. Involvement of such carriers in Cd uptake by several other cell types, however, repeatedly has been suggested (see section 6-8), although at times without consideration of the alternative mechanism proposed for the rat jejunum.

Following their apical uptake into enterocytes, several heavy metals become trapped and may be retained in the mucosa for the normal lifespan of the cells. Subsequent desquamation of the cells then leads to their loss from the body; this constitutes an important mechanism for protecting the body against excessive metal uptake. A cell constituent significantly involved in metal sequestration by enterocytes (and other cells) is cytoplasmic metallothionein (Foulkes and McMullen,1986). Cadmium metallothionein does not readily cross most normal cell membranes. Thus, it is not readily absorbed from the lumen of the perfused intestine. That some absorption of Cd ligated to metallothionein or a breakdown product does occur is indicated, however, by comparison of Cd distribution in the body following intestinal exposure to either $CdCl_2$ or Cd-metallothionein (Cherian, Goyer, Valberg, 1978). An exception to the generally small cellular uptake of Cd-metallothionein is found at the brush border of the proximal renal tubule where a carrier system has been identified specific for low–molecular weight anionic proteins such as metallothionein (Foulkes, 1978). This observation can explain the ability of the kidney to take up Cd-metallothionein released from the liver.

Unless heavy metals transferred from the intestinal lumen across the apical membrane into enteric epithelial cells are sequestered by compounds like metallothionein, which normally does not readily diffuse through cell membranes, they may be extruded across the basolateral membrane. This extrusion appears to involve the metal in the form of low–molecular weight diffusible complexes with ligands such as glutathione or cysteine (Foulkes, 1993). The suggestion that metals diffuse passively out of cells as complexes with, for example, glutathione, rather than following a reaction with membrane constituents as they do during apical uptake, is supported by the observation that Cd extrusion remains unaffected by even relatively high intracellular Zn concentrations (Foulkes and McMullen, 1986a). If the process required binding of the free metal to the basolateral membrane, as demonstrated for the apical membrane, then one could expect an inhibitory effect of excess Zn. Similar results were obtained for step 2 of Ni absorption in the jejunum (Foulkes and McMullen, 1986b).

The hypothesis that extrusion of heavy metals across the cell membrane involves formation of diffusible complexes was specifically tested by analyzing the effects of exogenous glutathione on the release of Cd and Hg from enteric cells (Foulkes, 1993). The results are summarized in Table 6-1.

The results clearly show that Hg but not Cd was mobilized by GSH. Such a result can be predicted on the basis of the respective affinities of these metals for metallothionein and glutathione. Metallothionein binds Cd very specifically and much more strongly than does glutathione. In contrast, binding of Hg to MT involves little more than its general affinity for sulfhydryl groups. As a result, excess glutathione can readily dissociate Hg from Hg-metallothionein but not Cd from Cd-metallothionein. Competition for intracellular metals between metallothionein and more diffusible sulfhydryl

TABLE 6-1. *Effect of glutathione on washout of Cd and Hg from segments of rat jejunum perfused in vivo*

	Cadmium, *nmol*		Mercury, *nmol*	
	Control	Glutathione	Control	Glutathione
Net removal from lumen during preperfusion[a]	28		19	
Recovered in tissue (after washout)	19	19	14	10
Recovered in luminal washout	9	9	5	9
Fractional washout of original tissue metal	32%	32%	26%	47%

[a]Net removal from lumen during 10 minutes preperfusion with saline containing 20 μM $CdCl_2$ effect on Hg washout significant at <.01 and $HgCl_2$ was equated to tissue content before washout (10 minutes with saline ±0.1 mM GSH).

Data from Foulkes (1993).

ligands like glutathione thus can explain the relatively ready washout of Hg, presumably as a glutathione complex. The same explanation is offered for the fact that, in spite of the much higher affinity of Hg than of Cd for metallothionein, Hg is retained in renal cortex for only weeks as contrasted to a half-life of Cd on the order of 30 years.

Although Cd washout from enteric cells is not affected by glutathione (see Table 6-1), monothiols depress apical Cd uptake from the lumen of saline-perfused jejunum. The most likely explanation for this finding is that the low–molecular weight complexing agent competes for Cd with the binding site on the cell membrane involved in step 1A of absorption and thus reduces Cd uptake. Very tight complexing of the metal as with EDTA, as pointed out previously, may abolish completely its uptake. Indeed, Cd-EDTA is an entirely inert complex whose renal clearance can serve as measure of glomerular filtration rate, and its distribution in the body remains entirely extracellular.

In spite of the suggested need for Cd complexes to dissociate prior to their absorption, some heavy-metal complexes in the intestine may be taken up to some extent at the jejunal brush border, either with minimal alteration or in unaltered form. This was deduced for Cd-metallothionein, as mentioned previously, from the finding that its slow jejunal absorption leads to the same distribution of absorbed Cd in the body as is observed after systemic administration of Cd-metallothionein; the metal in either case is accumulated primarily in the kidney rather than the liver (Cherian, Goyer, and Valberg, 1978). In contrast, orally or intravenously administered inorganic Cd mostly is recovered from the liver; this presumably reflects the very rapid binding of Cd to plasma protein (Foulkes, 1974) and its consequent uptake in the liver.

Monomethyl mercury (MMHg) can be absorbed from the intestine as a cysteine complex with the mediation of an amino acid transport system (Hirayama, 1975). A similar process has been described for the blood–brain barrier, as further discussed in section 6-6. In addition to the greater lipid solubility of MMHg than of inorganic mercury, such carrier mediation helps explain both the ready absorption of the compound

TABLE 6-2. Intrinsic permeability of jejunal mucosa: Physiological variation and in vitro artifact

	Flux ratio solute/ethanol			
	Urea	Cd	Hg	Cycloleucine
Mature rats	0.18	0.60	0.39	0.67
Weanling rats	0.43	0.94	0.85	0.62
Everted sacs (mature rats)	0.75	—	—	—

The unidirectional rate constant of passive ethanol efflux from the lumen at a given temperature is a function of the size, but not the composition, of the absorbing area. The ratio of solute to ethanol efflux rates therefore provides in arbitrary units a measure of the intrinsic permeability of the mucosal membrane to this solute. Data from Foulkes and Bergman (1993).

from the intestinal lumen and its rapid access to the central nervous system. Inorganic metals, in contrast, generally are absorbed from the mature intestine to the fractional extent of only a few percent.

The permeability of mucosal cell membranes to metals and other toxicants does not remain constant under various conditions. Physical handling of a tissue (see section 2-3), physiologic factors such as diet and age, presence of chelators such as EDTA, and potentially the action of toxicants themselves all can alter the so-called intrinsic solute permeability of cell barriers. This measure of permeability to a given solute is defined as the unidirectional flux of this solute across the membrane divided by the membrane surface area involved. This area may be determined morphometrically, but its estimation especially for convoluted structures in vivo presents difficulties. A useful functional approach to this problem is based on the passive flux of a compound like ethanol, freely soluble in both lipids and water. The result, as discussed in sections 3-5 and 6-2, is a flux ratio expressing the intrinsic permeability in empirical units.

Table 6-2 documents how application of this technique led to the conclusion, for instance, that maturation of the intestine in the rat is accompanied by a 58% drop in its intrinsic permeability for urea, and 54% for Hg^{++}; the relative specificity of these developmental changes is shown by the fact that the permeability for cycloleucine does not decrease with age. The intrinsic permeability to urea here provides a measure for the contribution of aqueous or polar pores to the total membrane area (Macey, 1984), a potentially important determinant of the intrinsic permeability of a membrane to nonlipophilic solutes (Foulkes and Bergman, 1993).

Note, as described in section 3-5, that increased urea permeability of jejunum is associated with the preparation of everted sacs. It is not known whether the fall in metal permeability with maturation reflects the loss of polar pores indicated by the fall in the urea/ethanol flux ratio. Characteristics of Cd uptake by enteric cells, however, are on the whole not compatible with involvement of such pores. More likely, the decreased metal uptake reflects the observed fall in membrane fluidity during development (Foulkes, 1988). Amino acid absorption, on the other hand, known to be mediated by specific amino acid transporters, showed no decrease with development (see Table 6-2). In other words, under these particular conditions the concentration of amino acid carriers on the membrane must have remained constant. This concentration too can change, however, by synthesis or recruitment of carrier molecules, as for instance during starvation (see Chapter 7).

6-4 PULMONARY TOXICANT ABSORPTION

Pulmonary epithelium is generally more solute-permeable than is intestinal epithelium. As a result, fractional absorption of an inhaled toxicant may greatly exceed its fractional absorption from the intestine. For instance, fecal excretion of ingested Cd averages 95% or more; only about 5% of the Cd load is retained by enteric cells or absorbed into the body, depending on the composition of intestinal contents and other factors. Much more complete fractional uptake of Cd has been observed in alveoli, amounting to 50% or higher (Friberg et al., 1985). The reasons why alveolar epithelium is so much more permeable to metals than is the enteric mucosa have not been established clearly. Possible explanations include, for instance, a greater fluidity of alveolar than of enteric cell membranes, and the relative absence from the alveolar space of compounds competing with membrane-binding sites for the metal.

Another possibility is that the alveolar membranes may contain a higher frequency of polar pores, such as found in the neonatal gut (see previously), or in everted sacs of intestine (see section 3-5). Nor has the role of tight junctions been compared in alveolar and intestinal epithelium: Perhaps intercellular diffusion is more significant in lungs than in intestine. Finally, the abilities of alveolar and enteric cells to trap heavy metals have not been compared yet.

6-5 HEPATIC TOXICANT TRANSPORT

Many xenobiotics are taken up by the liver, a major target organ for toxic agents. There they may be metabolized to more- or less-toxic compounds, lead to cytotoxic effects, and be released into plasma or secreted into bile. Secretion into bile provides for some solutes a major pathway of excretion. Both uptake and excretion involve solute transfer across membranes, so that cytotoxicity itself may alter uptake and excretion of a xenobiotic.

Such a mechanism has been proposed for the production of renal damage by Cd. The metal first is taken up largely by the liver, where it is to a great extent sequestered as the metallothionein complex. Judging from the fact that infusion of Cd-metallothionein, even at very slow rates not causing major increases in its plasma concentration (see Table 6-3), leads to renal rather than hepatic accumulation of Cd, hepatic cell membranes do not react with Cd-metallothionein as effectively as does the brush border in the renal tubule. Other cell types investigated also react only slowly with Cd-metallothionein (Borghesi and Lynes, 1996). The metallothionein induced in liver by Cd sequesters the metal and prevents its continued biliary excretion. Damage to the liver, however, whether related to its Cd load or produced by toxicants like carbon tetrachloride, releases Cd-metallothionein into plasma. The end result is renal accumulation of the compound and renal intoxication.

TABLE 6-3. *Renal and hepatic Cd accumulation after long-term slow infusion of CdCl₂ and CdMT*

	Cadmium chloride			Cadmium metallothionein		
	Liver	Kidney (cortex)	K/L	Liver	Kidney (cortex)	K/L
Tissue Cd, nmol/g wet weight	3.6	3.9	1.1	2.5	10.5	4.2
Total Cd, % of infused dose	22	4	0.2	14	12	0.9

Cd was infused with an osmotic minipump over a period of 72 hours at a rate of 2.8 nmol/h into adult rats (average weight 250 g); plasma Cd levels were estimated to have approximated 7 ng/dL. K/L values represent ratios.
Data from Johnson and Foulkes (1980).

Mechanisms of heavy-metal uptake also have been analyzed repeatedly in isolated hepatocytes. The findings in many ways resemble those made with enterocytes, although the suggested interpretations differ greatly. In particular, participation of energy-requiring transport in hepatic uptake has been suggested on the basis of its temperature dependence, but this hypothesis is not adequately supported by the reported findings (see section 6-2).

A role of Ca channels in Cd uptake also has been suggested for hepatocytes (Blazka and Shaikh, 1991); however, the evidence for competition between Ca and Cd at the membrane in these cells is not complete. For instance, in the work of Souza et al. (1996) with human fetal hepatic cells, competition between the two metals was proposed even though raising the Cd concentration from 10 to 500 μM did not significantly increase the fractional inhibition of Ca transport; this remained near 40%. It was precisely such lack of correlation between fractional inhibition and inhibitor concentration that led to the rejection of the hypothesis of competition between Ca and Cd in the enterocyte (Foulkes, 1985).

The possible role of Ca channels proposed for the mechanism of Cd uptake by hepatocytes and other cells is further discussed in section 6-8. Strong support for such a hypothesis is found, for instance, in the work of Hinkle, Kinsella, and Osterhoudt (1987) with secretory cells from the pituitary. There is some evidence also for the participation of Ca channels in Cd uptake in other cells, but the issue remains unresolved. Similarly, a participation of Ca channels in intestinal Cd uptake remains under discussion (see section 6-3).

Judging from the small fractional uptake of exogenous Cd-metallothionein by the liver at even low plasma concentrations (see Table 6-3), the cell membranes of hepatocytes do not appear to react appreciably with Cd-metallothionein. Hepatocyte membranes thus provide a measure of protection against plasma Cd-metallothionein. Evidence that Cd-metallothionein interferes with membrane function (Cherian, 1985) need not indicate that the complex exerts a direct effect on the membrane; the finding also can be interpreted in terms of an indirect inhibition (see Chapter 5).

In contrast to the protective role of hepatocyte membranes against Cd-metallothionein, specific membrane carriers appear to be responsible for the high hepatotoxicity of microcystins. These are cyclic peptide toxins produced by cyanobacteria (Runnegar, Berndt, and Kaplowitz, 1995), and their toxicity is correlated with the rate of their transport across the cell membranes. The membrane transport system involved differs from the well-characterized bile-salt organic anion transporter. Uptake is depressed by a miscellaneous group of chemoprotectants, including the antibiotic rifampin and the immunosuppressant cyclosporin A.

6-6 RENAL TOXICANT TRANSPORT

The kidneys provide the major pathway of excretion for many xenobiotics or their conjugates and metabolites. This excretion is controlled by the permeability of the glomerular filtration barrier and the ability of tubular cell membranes to reabsorb or secrete solutes. For instance, the effective renal excretion of water-soluble detoxification products such as glucuronides, mercapturic acids, or glycine conjugates is based on their tubular secretion and the absence of significant tubular reabsorption. Because the topic of renal excretion of xenobiotics has been extensively reviewed in the past, it is illustrated here only by some selected examples in which tubular cell membranes clearly are involved.

Any interference with normal membrane function in the kidney is reflected, of course, in renal malfunction. This is true of the therapeutic effects of diuretics as well as the cytotoxic actions of heavy metals. In addition, renal membrane transport systems also contribute to the cytotoxic potential of several nephrotoxicants by mediating their accumulation at intracellular target sites. Renal accumulation and retention of xenobiotics, indeed, contribute to the great susceptibility of the kidneys to intoxication.

A well-studied example of renal toxicant accumulation is that of heavy metals like Cd or Hg. As already referred to in section 6-5, the presence of metallothionein transporters specifically in the renal brush border presumably accounts for the higher renal than hepatic accumulation of Cd circulating as Cd-metallothionein. Such preferential renal uptake of Cd-metallothionein was demonstrated in studies in which the protein was adminstered systemically with an osmotic minipump at a rate approaching that of the likely hepatic release of Cd-metallothionein (Johnson and Foulkes, 1980); Table 6-3 summarizes results of these studies. Cadmium-labeled Cd-metallothionein or $CdCl_2$ was infused for a period of 72 hours; the animals then were killed and Cd levels determined in liver and kidney.

Renal accumulation of Cd-metallothionein, as already stated, is mediated by an apical transport system in the proximal tubule specific for anionic proteins like metallothionein or myoglobin (Foulkes, 1978). Unlike apical membranes, basolateral membranes do not react with metallothionein (Foulkes, 1991b). They readily clear, however, metals like Cd, Zn and Hg from plasma, provided that these metals circulate in a diffusible form (Foulkes, 1974) rather than complexed to plasma protein, and that binding sites on the cell membranes can compete for the metals with diffusible ligands in plasma or extracellular fluid, as discussed in section 6-3. The ability of the kidney to accumulate Cd circulating in plasma complexed to, for example, cysteine or glutathione raises questions about the assumption that Cd is transported to the kidney exclusively in the form of Cd-metallothionein, as implied, for instance, by Squibb (1996); demonstration of Cd-metallothionein in plasma following Cd exposure does not exclude the presence of these other Cd complexes.

The mechanism whereby Cd-metallothionein taken up from glomerular filtrate damages the tubule cells remains under discussion. A commonly suggested explanation attributes the toxic effects to Cd liberated from Cd-metallothionein in the lysosomes; at that stage the Cd is believed to act as does Cd taken up in the form of, for example, Cd-glutathione. The assumption that Cd from Cd-metallothionein and Cd-glutathione enters the same toxic metal pool, however, is untested. One could plausibly argue that lysosomal Cd from Cd-metallothionein primarily damages the lysosomal membrane, and that the consequent cytotoxic effects are caused not by Cd but by leakage of hydrolytic enzymes into the cytoplasm. In that case, there is no reason why Cd taken up as Cd-metallothionein should enter the same pool as that taken up in the form of, for example, Cd-glutathione; nor would one expect that the Cd from the two sources would exert identical effects. The implications of such a conclusion for the concept of a unique critical level of Cd in renal cortex are obvious (see, for example, Foulkes, 1990).

Reaction of heavy metals with basolateral cell membranes as part of their diffusion-limited uptake in the renal cortex is demonstrated readily by a modification of the rapid-transit technique in the living animal (Foulkes, 1989). In these experiments, the metal was injected in an arterial bolus together with a glomerular and extracellular marker such as inulin. The injected bolus was trapped in the kidney by temporary arterial occlusion (40 seconds). After resumption of normal perfusion, the venous recovery of the metal was compared with that of inulin; a deficit in the recovery of metal indicates its removal from the postglomerular extracellular space. Figure 6-3 illustrates the results of such an experiment on basolateral uptake of Cd in the rabbit kidney.

Note that during a single passage through the kidney, without occlusion, venous recovery of Cd resembles that of inulin; the slight excess of Cd probably reflects less-than-complete filterability of the metal because of its binding to plasma protein, in spite of the presence of a low–molecular weight thiol (mercaptoethanol). In contrast, on the right, the diffusion-limited process is able during occlusion to proceed toward completion; as a result approximately one third of the injected Cd is extracted in the kidney.

If bound to plasma proteins, metals essentially become nondiffusible and nonfilterable, unable to reach the renal parenchyma (Foulkes, 1974). In other words, the metals must be bound in diffusible complexes as otherwise they would neither be filtered at the glomerulus nor be able to diffuse to and react with basolateral cell membranes. Even diffusible metals, however, are prevented from reacting with cell membranes if they are very tightly, and to that extent essentially irreversibly, ligated. As a result, the EDTA complexes of Cr and Cd, for instance, are biologically inert and are handled by kidneys and intestine like inulin (see section 6-3). These findings are documented for the renal handling of Cd in Table 6-4.

TABLE 6-4. Renal handling of cadmium following arterial injection

	Recoveries %			
	Vein	Urine	Total	Transit time, s
Inulin	75	25	100	10.0
Cd (as CdCl$_2$)	99	0	99	8.4
Cd EDTA	70	20	90	—
Cd mercaptoethanol	63	4	67	10.0[a]

[a]The identical values for inulin and CdME were described in Foulkes (1989) and Foulkes and Blanck (1990).
Arterial boluses containing Cd and inulin were injected into rabbits. Mean transit times were measured from artery to vein. Recoveries were calculated on the basis of the value for inulin of 100%.
Data from Foulkes (1974).

Injection of CdCl$_2$ here leads to the essentially complete recovery of Cd in the renal vein, after a mean transit time shorter than that of inulin; this presumably reflects the binding of the metal to plasma proteins whose vascular transit time across the kidney is shorter than that of nonfiltered inulin. Because none of the Cd was filtered or retained in the kidney under these conditions, its venous recovery exceeds that of inulin.

In presence of EDTA, the metal behaves like inulin, with the same distribution between venous plasma and urine, and again no retention. In contrast, if injected with

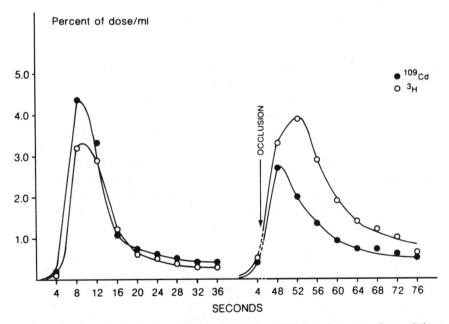

FIG. 6-3. Cd uptake at the basolateral cell membrane in rabbit kidney. Artery-to-vein transit times across the kidney are shown for Cd and inulin ion control (*left*) and with 40 seconds of vascular occlusion (*right*). (Reprinted with permission from Foulkes, 1989.)

an excess of mercaptoethanol, Cd again was observed to become freely diffusible, as indicated by a mean artery-to-vein transit time equal to that of inulin. However, mercaptoethanol does not possess sufficient affinity for Cd to be able to compete with Cd-binding sites on renal cell membranes. The low recovery of filtered Cd in urine in these experiments therefore must result from tubular reabsorption. Cd also is taken up from the postglomerular plasma, as shown by its lower venous recovery than that of inulin.

As discussed previously, and in contrast to Cd, Hg is not bound by EDTA as strongly as it is to binding sites on the cell membranes (Foulkes and Bergman, 1993). This observation suggests a role of sulfhydryl groups in Hg binding, based on the fact that the affinity of Hg^{++} for -SH greatly exceeds that for EDTA. As a result, in neither kidney nor intestine can EDTA prevent the reaction of Hg with the cell membrane (Foulkes, 1991a).

Carrier-mediated transport of Cd has been suggested for rat renal cortical cells (Endo, Sakata, and Shaikh, 1995). One of the main arguments adduced in support of this interpretation was the observed temperature-dependence of the process. This, however, cannot be equated to a requirement for metabolic energy. Indeed, the Q_{10} value for this process can be calculated to equal less than 1.5, little more than the temperature-dependence of simple physical diffusion or the viscosity of water ($Q_{10} = 1.2$). The explanation offered in section 6-3 for the same Q_{10} value for Cd uptake by enteric cells invokes changes in membrane fluidity. The evidence for energy-requiring Cd uptake by renal cells thus remains unconvincing.

An important aspect of the role of renal membranes in toxicokinetics is the transport of selected toxicants across the basolateral and apical cell membranes in the direction of either tubular absorption or secretion. For instance, the major fraction of mercury excreted in urine following the intravenous injection of chlormerodrin into dogs appears to be contributed by tubular secretion (Borghgraef, Kessler, and Pitts, 1956). Another example is the renal handling of the mycotoxin ochratoxin, which is reabsorbed in the rat kidney (Zingerle, Silbernagl, and Gekle, 1997) and secreted across the tubular epithelium in nonfiltering toad kidneys (Bahnemann et al., 1997). Chronic exposure to the agent inhibits organic anion transport at the basolateral cell membrane of OK cells (Sauvant, Silbernagl, and Gekle, 1997). In addition to being transported across membranes, the toxin therefore also can inhibit membrane function (see also Gekle, Oberleithner, and Silbernagl, 1993). Whether these effects are directly or indirectly exerted on the membrane (see Chapter 5) is not known.

An instance of specifically basolateral uptake of a toxicant by renal tubule cells is that of paraquat, as demonstrated by Groves et al. (1995). This process is mediated by a carrier system specific for polyvalent organic cations, and its action helps explain the nephrotoxicity of the compound. A similar active toxicant accumulation is mediated by the organic anion-transport system at the basolateral membrane. This reacts with beta-lactam antibiotics like cephalosporin (Tune, 1988), which as a result reach their potential intracellular target site at significant concentrations. The toxicologic significance of the finding is emphasized by further evidence: Only those lactams that undergo tubular secretion produce toxic injury; this injury is restricted to the proximal tubule, the localization of the organic anion transport system; toxic thresholds correlate with the cortical concentration of these compounds; and toxicity is prevented by agents like Probenecid, the well-known inhibitor of tubular anion transport.

Similarly, the nephrotoxic action of cisplatin depends on membrane transport mechanisms (Borch, 1993); both the so-called organic anion and cation carrier systems may be involved. The renal clearance of cisplatin, however, equals that of nonnephrotoxic analogues like carboplatin (Daley-Yates and McBrien, 1985). Membrane transport of cisplatin per se, therefore, cannot fully explain the nephrotoxic actions of the compound.

The nephrotoxicity of certain cysteine and glutathione conjugates also is based on the activity of membrane mechanisms mediating their specific uptake by the kidney (Lash and Anders, 1989). The toxic action itself involves the "lethal synthesis" in the cells of highly reactive S compounds that bind to cell constituents. Membrane transport of another potentially toxic compound was described by Silbernagl, O'Donovan, and Volker (1997), who found that the amino acid D-serine is transported by a D–amino acid carrier into collecting duct cells, where its oxidation by D–amino acid oxidase produces H_2O_2 and consequent oxidative stress. Even normal concentrations of D–amino acids, making up perhaps 1% of circulating amino acids, thus could constitute a significant toxicologic challenge.

Renal-cell membranes also contribute to the secretion of certain xenobiotics into the urine. Reference in this connection has been made to the tubular secretion of chlormerodrin (Borghgraef, Kessler, and Pitts, 1956). A further contribution of membrane transport to urinary excretion is discussed in Chapter 7; it consists of the action of a broadly specific membrane pump involved in the active cellular extrusion of xenobiotics and consequently responsible for multiple drug resistance of many cells. A question that has not been clarified fully is whether extrusion of a toxicant or other solute from cells during its transcellular movement is mediated by the same mechanism as that responsible for its contralateral uptake into the cell. The two processes appear to differ for heavy-metal absorption in the intestine (section 6-3). On the other hand, renal tubular amino acid reabsorption involves very similar transport systems on the apical and basolateral cell membranes, both sensitive to heavy metals and showing similar substrate specificities (Foulkes, 1987).

This section, in summary, emphasizes the important roles renal membrane mechanisms play as targets of cytotoxic action, as mediators of cytotoxicity, and as determinants of xenobiotic excretion.

6-7 BLOOD–BRAIN BARRIER

Solute access to interstitial spaces in the central nervous system is well known to be limited by the capillary endothelium acting as blood–brain barrier. A function of this barrier is to control the composition of the extracellular fluid in the central nervous system, and thus to protect the brain against xenobiotics circulating in blood. A well-known example of the effectiveness of the barrier in fulfilling this function is provided by the different toxicities of inorganic mercury, organic mercury, and mercury vapor. The blood–brain barrier cannot prevent entrance of highly lipid-soluble dimethyl mercury or mercury vapor into the central nervous system. In contrast, inorganic mercury essentially is excluded from the central nervous system and instead accumulates in the kidney as its main target organ. A similar determinant role of lipid solubility in tranfer of Hg compounds across cell membranes was demonstrated in myocardium (Halbach, 1990). Here, signs of intoxication after exposure appeared only after an initiation delay inversely proportional to the lipid solubility of the compounds tested.

The lipid solubility of MMHg (labeled MeHg in Fig. 6-4) lies between those of Hg^{++} and dimethyl mercury. MMHg is a monovalent cation, and as such its ready movement across the blood–brain barrier cannot be fully accounted for by its solubility properties. It is interesting, therefore, to find that the cerebral uptake of this compound also depends on the presence of membrane carrier mechanisms in the blood–brain barrier. Transfer of MMHg across this barrier involves the cysteine complex (Kerper, Ballatori, and Clarkson, 1992), and is believed to be mediated by an amino acid carrier system. The findings were made with an indicator dilution technique similar to that described for renal solute uptake in section 6-6 and are illustrated in Fig. 6-4; they recall the work on intestinal transport of MMHg discussed in section 6-3. Uptake of MMHg by brain was calculated from the ratios of MMHg to extracellular indicator in the bolus and in venous plasma collected 15 seconds after arterial injection. It is worth emphasizing that the double-indicator dilution technique cannot distinguish between retention in the blood–brain barrier and accumulation in the brain itself.

Two important points emerge from these results. First, note that only the complex containing L-cysteine is taken up at a significant rate; the complex with D-cysteine

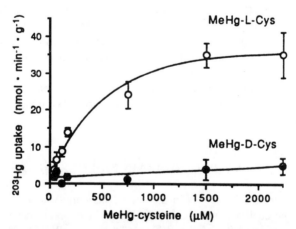

FIG. 6-4. Monomethylmercury transport across the blood–brain barrier in rats in presence of cysteine. (From Kerper, Ballatori, and Clarkson, 1992; with permission.)

is essentially inert. Secondly, the process appears to be saturable. Further, transport of (L-cysteine)-MMHg was found to be depressed, presumably competitively, by L-methionine, a typical substrate for the L-type carrier system for neutral amino acids. The A-type substrate methylaminoisobutyrate was inactive.

Not only cellular uptake of MMHg but also its extrusion may be mediated by membrane carrier systems (Fujiyama, Hirayamo, and Yasutake, 1994). Specifically, efflux of the mercurial from astrocytes in culture has been reported to follow its conjugation with glutathione. Extrusion of the mercurial from the cells is depressed by Probenecid, the classical inhibitor of the organic anion transport system. It is worth recalling, in this connection, the conclusion that glutathione complexes may be the form in which heavy metals generally are extruded across cell membranes (see section 6-2), although no evidence has been reported of specific mediation of this step in cells other than astrocytes.

Although a major function of the blood–brain barrier is thus to prevent passage of nonlipophilic toxic agents into the central nervous system, the barrier cannot prevent completely uptake of such agents by the brain. An instance of toxicant penetration through the barrier was reported by Kerper and Hinkle (1997). These authors described uptake of Pb by brain capillary endothelium. This uptake is activated by depletion of intracellular Ca^{++} stores, suggesting a possible role of store-gated cation channels in Pb uptake.

The barrier itself may become a target of toxic action. As pointed out by Romero, Abbott, and Bradbury (1996) in special reference to the metals lead, mercury, aluminum, iron, and cadmium, toxic changes in the permeability of the barrier have been observed frequently. They may result from at least three different processes: 1) direct action on the integrity of the barrier, which could increase its permeability, with consequent penetration of plasma solutes and water into the central nervous system and the production of cerebral edema; 2) alterations of specific membrane transport systems, as illustrated by the ability of 10 μM lead to inhibit glucose uptake by mouse brain endothelial cells (Maxwell et al., 1986), which could depress the supply of essential metabolic substrates; and 3) a direct action on neuronal or glial function affecting cerebral metabolism, which can alter indirectly the functional integrity of the barrier.

It is relevant here to mention certain similarities in the toxicologic functions of the blood–brain barrier and of the placenta. The placental barrier protects the fetus by controlling tranfer of toxic agents from mother to fetus. It also serves, however, as conduit for heavy metals and may itself become a target for these metals. Cadmium, for instance, depresses placental transport of cyanocobalamine, amino acids, Cu, Fe, and Zn (for review, see Eisenmann and Miller, 1996).

Although the protective role of the blood–brain barrier, as well as the contribution of membrane transport to transfer across this barrier, are discussed here primarily in terms of metal compounds, access of many other toxicants and drugs to the central nervous system is, of course, well known to be similarly controlled by the properties of cell membranes in the barrier. The role of the multidrug exporting transporter in controlling access of xenobiotics to the central nervous system is considered in section 7-3.

6-8 MEMBRANE TRANSPORT OF TOXICANTS IN OTHER CELL TYPES

Membranes of all cells studied potentially participate in controlling uptake and distribution of nonlipid soluble xenobiotics. Thus, dermal absorption of xenobiotics is a function of the permeability of cell membranes. Even in procaryotic cells, sensitivity to drugs often is determined at the cell membrane. This has been observed, for instance, by Jude et al. (1996), who reported that wild type *Esherichia coli* is sensitive to the antibacterial agent shikimic acid and accumulates it from the medium. In contrast, no uptake is seen by a shikimate-resistant mutant.

In general, different cell types frequently are found to possess distinctive mechanisms for transferring a given toxicant across the cell membrane. A prime example of multiple mechanisms for transporting a single toxicant is that of Cd uptake. Thus, section 6-3 describes a model that adequately explains transfer of this metal across apical cell membranes in the jejunal mucosa of the rat. The model, however, may not apply to all other cell types. Unlike in the jejunal mucosa, secretory cells from the pituitary, for instance, have been reported by Hinkle, Kinsella, and Osterhoudt (1987) to accumulate Cd through voltage-gated Ca channels; this clearly differs from metal absorption in rat jejunum, in which no competition between Cd and Ca can be observed (see section 6-3).

Metal uptake by erythrocytes also differs in several regards from that observed in jejunum. The finding that the process is inhibited by N-ethylmaleamide suggested to Garty, Bracken, and Klaassen (1986) that reactive sulfhydryl groups are essential; no evidence for such a role for sulfhydryl compounds was obtained in jejunum (Foulkes, 1991a). The erythrocyte membrane also differs from other membranes in its very high anion permeability. Anion channels appear to be significantly involved in uptake of several heavy metals by erythrocytes, as deduced from the effect of the anion channel inhibitor DIDS (see section 6-3). The explanation presumably lies in formation of anionic polychloride complexes of heavy metals in physiologic saline, especially in the case of Hg^{++}. It must be recalled, however, that DIDS exerts no effect on Hg uptake in jejunum (Foulkes and Bergman, 1993), even though evidence strongly suggests that here too the metal exists in the saline perfusate largely in anionic form, and that it first reacts with the membrane as an anion. The fact that DIDS inhibits Hg uptake by erythrocytes but does not alter that by enteric cells suggests that step 1B of Hg absorption in the intestine (see section 6-3) does not involve the metal in an anionic form.

6-9 CONCLUSIONS

The main point emphasized in this chapter is the well-established fact that not only can cell membranes themselves serve as targets of toxic action, but they also are important determinants of the toxicokinetic characteristics of nonlipophilic xenobiotics. Indeed, the intrinsic permeability of cell membranes, defined by unidirectional transmembrane fluxes across unit surface area, largely accounts for uptake, distribution, and excretion of xenobiotics. Wide variations are found in the intrinsic permeability properties of cells from different tissues, as could have been predicted from the organ-specific distribution and toxicity of many xenobiotics. For instance, plasma cadmium metallothionein is taken up in the kidney but not in the liver. A similar explanation may account for the preferential renal uptake of inorganic mercury.

Even if a xenobiotic is transported across cell membranes, however, the mechanisms responsible may vary significantly from one cell type to another. The existence of such multiple mechanisms for transporting a given toxicant across cell membranes has been discussed specifically in relation to the uptake of cadmium by different cell types.

Strong accumulation and retention of a toxicant in a specific tissue, of course, also may reflect intracellular trapping mechanisms, as in the case of metallothionein and heavy metals. The sensitivity of a tissue to a toxicant also depends on the possible presence of membrane transport mechanisms actively mediating its extrusion from the cells. Such extrusion increases the cellular resistance to the agent, as further discussed in the next chapter. As an overall conclusion we may state that the net rate of uptake of a toxicant, the organ specificity of its distribution, the concentration it can reach in the target organ, its biologic half life, and the rate of its excretion all are determined to a large extent by the properties of cell membranes.

REFERENCES

Aungst, B. 1996. Oral mucosal permeation enhancement. In *Oral mucosal drug delivery*, ed. M.J. Rathbone, 65–83. New York: Marcel Dekker.

Bahnemann, E., H.P. Kerling, G. Ensminger, G. Schwerdt, S. Silbernagl, and M. Gekle. 1997. Renal transepithelial secretion of ochratoxin A in the non-filtering toad kidney. *Toxicology* 120:11–17.

Bennett, E., M.S. Urcan, S.S. Tinkle, S.R. Koszowski, and S.R. Levinson. 1997. Contribution of sialic acid to the voltage dependence of sodium channel gating. *J. Gen. Physiol.* 109:327–343.

Bevan, C., and E.C. Foulkes. 1989. Interaction of cadmium with brush border membrane vesicles from the rat small intestine. *Toxicology* 54:297–309.

Bevan, C., E. Kinne-Saffran, E.C. Foulkes, and R.K.H. Kinne. 1989. Cadmium inhibition of L-alanine transport into renal brush border membrane vesicles isolated from the winter flounder. *Toxicol. Appl. Pharmacol.* 101:461–469.

Blazka, M.E., and Z.A. Shaikh. 1991. Differences in cadmium and mercury uptake by hepatocytes: a role of calcium channels. *Toxicol. Appl. Pharmacol.* 110:355–363.

Borch, K.F. 1993. The nephrotoxicity of anticancer drugs. In *Renal disposition and nephrotoxicity of xenobiotics*, eds. M.W. Anders, W. Dekant, D. Henschler, H. Oberleithner, and S. Silbernagl, 289–310. New York: Academic Press.

Borghesi, L.A., and M.A. Lynes. 1996. Non-protective effects of extracellular metallothionein. *Toxicol. Appl. Pharmacol.* 139:6–14.

Borghgraef, R.R.M., R.H. Kessler, and R.F. Pitts. 1956. Plasma regression, distribution and excretion of radio-mercury in relation to diuresis following the intravenous administration of ^{203}Hg labelled chormerodrin to the dog. *J. Clin. Invest.* 35:1055–1066.

Cherian, M.G. 1985. Rat kidney epithelial cell culture for metal toxicity studies. *In Vitro Cell Dev. Biol.* 21:505–508.

Cherian, M.G., R.A. Goyer, and L.S. Valberg. 1978. Gastrointestinal absorption and organ distribution of oral $CdCl_2$ and Cd-metallothionein in mice. *J. Toxicol. Environ. Health* 4:861–868.

Cotzias, G.C., D.C. Borg, and B. Selleck. 1961. Virtual absence of turnover in cadmium metabolism: ^{109}Cd studies in the mouse. *Am. J. Physiol.* 201:927–930.

Csaky, T.Z. 1984. *Pharmacology of Intestinal Permiation*, Vol. 70, *Handbook of Experimental Pharmacology*. Berlin:Springer–Verlag.

Daley-Yates, P.T., and D.C.H. McBrien. 1985. The renal fractional clearance of platinum antitumor compounds in relation to nephrotoxicity. *Biochem. Pharmacol.* 34:1423–1428.

Eisenmann, C.J., and R.K. Miller. 1996. Placental transport, metabolism and toxicity of metals. In *Toxicology of metals*, ed. L.W. Chang, 1003–1026. Boca Raton, FL: Lewis.

Endo, T., M. Sakata, and Z.A. Shaikh. 1995. Mercury uptake by primary cultures of rat renal cortical epithelial cells: I. Effects of cell density, temperature and metabolic inhibitors. *Toxicol. Appl. Pharmacol.* 132:36–43.

Foulkes, E.C. 1974. Excretion and retention of cadmium, zinc and mercury by the rabbit kidney. *Am. J. Physiol.* 227:1356–1360.

Foulkes, E.C. 1978. Renal tubular transport of cadmium metallothionein. *Toxicol. Appl. Pharmacol.* 45:505–510.

Foulkes, E.C. 1985. Interactions between metals in rat jejunum: Implications on the nature of cadmium uptake. *Toxicology* 37:117–125.

Foulkes, E.C. 1987. Role of basolateral membranes in organic solute reabsorption in rabbit kidney. *Am. J. Physiol.* 252:F1042–F1047.

Foulkes, E.C. 1988. On the mechanism of transfer of heavy metals across cell membranes. *Toxicology* 52:263–272.

Foulkes, E.C. 1989. On the mechanism of cellular cadmium uptake. *Biol. Trace Element Res.* 21:195–200.

Foulkes, E.C. 1990. The concept of critical levels of toxic heavy metals in target tissues. *CRC Crit. Rev. Toxicol.* 20:327–339.

Foulkes, E.C. 1991a. Further findings on the mechanism of cadmium uptake by intestinal mucosal cells (step 1 of Cd absorption). *Toxicology* 70:261–270.

Foulkes, E.C. 1991b. Role of metallothionein in epithelial transport and sequestration of cadmium. In *Metallothionein in biology and medicine*, eds. C.D. Klaassen and K.T. Suzuki, 171–182, Boca Raton, FL: CRC Press.

Foulkes, E.C. 1993. Metallothionein and glutathione as determinants of cellular retention and extrusion of Cd and Hg. *Life Sci.* 52:1617–1620.

Foulkes, E.C. 1994. Epithelial transport of heavy metals. *Adv. Comp. Environ. Physiol.* 20:55–84.

Foulkes, E.C., and S. Blanck. 1990. Acute cadmium uptake by rabbit kidneys: Mechanisms and effects. *Toxicol. Appl. Pharmacol.* 102:464–473.

Foulkes, E.C., and D. Bergman. 1993. Inorganic Hg absorption in mature and immature rat jejunum: Transcellular and intercellular pathways in vivo and in everted sacs. *Toxicol. Appl. Pharmacol.* 120:89–95.

Foulkes, E.C., and D.M. McMullen. 1986a. Endogenous metallothionein as determinant of intestinal cadmium absorption: A reevaluation. *Toxicology* 38:285–291.

Foulkes, E.C., and D.M. McMullen. 1986b. On the mechanism of nickel absorption in the rat jejunum. *Toxicology* 38:35–42.

Foulkes, E.C., and D.M. McMullen. 1987. Kinetics of trans-epithelial movement of heavy metals in rat jejunum. *Am. J. Physiol.* 253:G134–G138.

Foulkes, E.C., and B.F. Miller. 1959. Steps in PAH transport by kidney slices. *Am. J. Physiol.* 196:86–92.

Foulkes, E.C., T. Mort, R. Buncher. 1991. Intestinal cadmium permeability in mature and immature rats. *Proc. Soc. Exp. Biol. Med.* 197:477–481.

Foulkes, E.C., and C. Voner. 1981. Effects of Zn status, bile and other endogenous factors on jejunal Cd absorption. *Toxicology* 22:115–122.

Friberg, L., C.-G. Elinder, T. Kjellstrom, and G.F. Nordberg. 1985. *Cadmium and health: A toxicological and epidemiological appraisal*. Boca Raton, FL: CRC Press.

Fujiyama, J., K. Hirayama, and A. Yasutake. 1994. Mechanism of methylmercury efflux from cultured astrocytes. *Biochem. Pharmacol.* 47:1525–1530

Garty, M., W.M. Bracken, and C.D. Klaassen. 1986. Cadmium uptake by rat red blood cells. *Toxicology* 42:111–119.

Gekle, M., H. Oberleithner, and S. Silbernagl. 1993. Ochratoxin A impairs post-proximal nephron function in vivo and blocks plasma membrane anion conductance in Madin-Darby canine kidney cells in vitro. *Pflugers Arch.* 425:401–408.

Groves, C.E., M.N. Morales, A.J. Gandolfi, W.W. Dantzler, and S.H. Wright. 1995. Peritubular paraquat transport in isolated renal proximal tubules. *J. Pharmacol. Exp. Ther.* 275:926–932.

Gunshin, H., B. Mackenzie, U.V. Berger, Y. Gunshin, M.F. Romero, W.F. Boron, S. Nussberger, J.L. Gollan, and M.A. Hediger. 1997. Cloning and characterization of a mammalian proton-coupled metal ion transporter. *Nature* 388:482–488.

Halbach, S. 1990. Mercury compounds: Lipophilicity and toxic effects on isolated myocardial tissue. *Arch. Toxicol.* 64:315–319.

Hart, B.A. 1978. Transport of cadmium by the alveolar macrophage. *J. Reticuloendoth. Soc.* 24:363–375.

Hinkle, P.M., P.A. Kinsella, and K.C. Osterhoudt. 1987. Cadmium uptake and toxicity via voltage-sensitive calcium channels. *J. Biol. Chem.* 262:16333–16337.

Hirayama, K. 1975. Transport mechanism of methyl mercury: Intestinal absorption, biliary excretion and distribution of methyl mercury. *Kumamoto Med. J.* 28:151–163.

Hoadley, J.E., and R.J. Cousins. 1985. Effects of dietary zinc depletion and food restriction on intestinal transport of cadmium in the rat. *Proc. Soc. Exp. Biol. Med.* 180:296–302.

Johnson, D.R., and E.C. Foulkes. 1980. On the proposed role of metallothionein in the transport of cadmium. *Environ. Res.* 21:360–365.

Jude, D.A., C.D.C. Ewart, G.L. Thain, G.M. Davies, and W.M. Nichols. 1996. Transport of the antibacterial agent (6S)-6-fluoroshikimate and other shikimate analogues by the shikimate transport system of E. coli. *Biochim. Biophys. Acta* 1279:125–129.

Kerper, L.E., N. Ballatori, and T.W. Clarkson. 1992. Methylmercury transport across the blood-brain barrier by an amino acid carrier. *Am. J. Physiol.* 262:R761–R765.

Kerper, L.E., and P.M. Hinkle. 1997. Lead uptake in brain capillary endothelial cells: Activation by Ca store depletion. *Toxicol. Appl. Pharmacol.* 146:127–133.

Kojima, S., and M. Kiyozumi. 1974. Studies on poisonous metals I: Transfer of cadmium chloride across rat small intestine in vitro and effect of chelating agents on its transfer. *Yakugaku Zasshi* 94:695–701.

Lash, L.H., and M.W. Anders. 1989. Uptake of nephrotoxic S-conjugates by isolated rat renal proximal tubular cells. *J. Pharmacol. Exp. Ther.* 248:531–537.

Lou, M., R. Garay, and J.O. Alda. 1991. Cadmium uptake through the anion exchanger in human red blood cells. *J. Physiol.* 443:123–136.

Macey, R.I. 1984. Transport of water and urea in red blood cells. *Am. J. Physiol.* 246:C195–C203.

Maxwell, K., H.V. Vinters, J.A. Berliner, J.V. Bready, and P.A. Cancilla. 1986. Effect of inorganic lead on some functions of the cerebral microvessel endothelium. *Toxicol. Appl. Pharmacol.* 84:389–399.

Nelson, D.J., L. Kiremidjian-Schumacher, and G. Stotzky. 1982. Effects of cadmium, lead and zinc on macrophage-mediated cytotoxicity toward tumor cells. *Environ. Res.* 28:154–163.

Paulsen, I.T., and M.H. Saier. 1997. A novel family of ubiquitous heavy metal ion transport proteins. *J. Membr. Biol.* 156:99–103.

Romero, I.A., N.J. Abbott, and M.W.B. Bradbury. 1996. The blood-brain barrier in normal CNS and in metal-induced neurotoxicity. In *Toxicity of metals*, ed. L.W. Chang, 561–585. Boca Raton, FL: Lewis Publishers.

Runnegar, M., N. Berndt, and N. Kaplowitz. 1995. Microcystin uptake and inhibition of protein phosphatases: Effects of chemoprotectants and self-inhibition in relation to known hepatic transporters. *Toxicol. Appl. Pharmacol.* 134:264–272.

Sandvig, K., and B. van Deurs. 1996. Endocytosis, intracellular transport, and cytotoxic action of Shiga Toxin and Ricin. *Physiol. Rev.* 76:949–966.

Sauvant, C., S. Silbernagl, and M. Gekle. 1997. Effect of chronic exposure to ochratoxin A on transport systems in OK cells [abstract]. *Pflugers Arch.* 433:R164.

Silbernagl, S., D.J. O'Donovan, and K. Volker. 1997. D-serine nephrotoxicity is mediated by oxidative damage. *Pflugers Archiv* 433 (supplement):R37.

Simons, T.J.B. 1986. The role of anion transport in the passive movement of lead across the human red cell membrane. *J. Physiol.* 443:123–136.

Solioz, M., and C. Vulpe. 1996. Cpx-type ATPases: A class of P-type ATPases that pump heavy metals. *Trends Biochem. Sci.* 21:237–241.

Souza, V., L. Bucio, D. Jay, E. Chavez, and M.C. Guiterrez-Ruiz. 1996. Effect of Cd on Ca transport in a human fetal hepatic cell line (WRL-68 cells). *Toxicology* 112:97–104.

Tune, B.M. 1988. The nephrotoxicity of cephalosporin antibiotics: Structure-activity relationships. *Commun. Toxicol.* 1:145–170.

Valberg, L.S., J. Sorbie, and D.L. Hamilton. 1976. Gastrointestinal metabolism of cadmium in experimental iron deficiency. *Am. J. Physiol.* 231:462–467.

Verbost, P.M., G. Flik, R.A. Lock, and S.E. Wendelaar-Bonga. 1988. Cadmium inhibits plasma membrane calcium transport. *J. Membr. Biol.* 102:97–104.

Zingerle, M., S. Silbernagl, and M. Gekle. 1997. Reabsorption of the nephrotoxin ochratoxin A along the rat nephron in vivo. *J. Pharmacol. Exp. Ther.* 280:220–224.

7

Extrusion of Toxicants and Metabolites from Cells and Adaptive Changes in Membrane Function

7-1 OVERVIEW

The preceding chapter discusses the transport of toxicants across cell membranes. Even in the absence of such transport, the permeability properties of membranes critically help determine the uptake, distribution, retention, and excretion of nonlipophilic toxic agents. In other words, as already considered in section 2-5, cell membranes play a determining role in toxicokinetics. This role is restricted primarily to nonlipophilic solutes because membranes offer little resistance to the passive movement of lipophilic substances.

Although mediated membrane transport of certain drugs or toxic agents actually may increase their cellular uptake (see Chapter 6), and thus raise inhibitor concentrations at intracellular target sites, more commonly membranes diminish net access to the cell interior. Two membrane properties primarily are responsible for this protective role against toxic xenobiotics. One of these, as repeatedly stated, is the impermeability of the membrane to selected toxicants. The second critical property of membranes in the present context is their constitutive or acquired ability to mediate in many instances the active extrusion of a wide range of toxic substances or their detoxification products from the cells.

The "pumping out" or export of toxicants from cells represents an instance of the broad ability of many cells to remove unwanted solutes and thus to maintain a suitable steady-state composition. Examples of such solute export are the excretion of waste products like lactic acid from working muscle, further discussed here, or the pumping out of sodium and consequent maintenance of low intracellular sodium concentrations in most cell types.

TABLE 7-1. *Apical and basolateral sensitivity of renal cycloleucine reabsorption to mercury*

	Control	Poisoned
Glomerular filtration, mL/min (SD)	5.0(0.4)	1.5(0.4)
Fractional reabsorption, % (SD)	>99	74(7)
Reabsorptive flux, nmol/min (SD)	100(8)	22(6)
Cortical cycloleucine, nmol (SD)	74(4)	210(74)
Cellular transit time, s (SD)		
Observed	40(4)	>120
Calculated	44	590

Mean steady-state values are shown with SD for the left kidney in six control rabbits and six rabbits that had received a subcutaneous injection of 10 μmol HgCl$_2$/kg body weight 48 hours prior to measurements. Plasma concentration of cycloleucine averaged 20 μM. The calculated cellular transit time is given by the quotient of tissue content/reabsorptive flux.
Data from Foulkes (1987).

Solute extrusion from cells is obviously also the final step in the transcellular movement of solutes across cell barriers, such as during renal tubular solute reabsorption or secretion or as part of intestinal absorption. The mechanism of such mediated extrusion of selected solutes in these instances may resemble that of the solute uptake process at the contralateral cell membrane, with similar substrate specificities and, in part, comparable sensitivities to inhibition by toxic agents. A basic difficulty in attempting to make such comparisons of inhibitor sensitivities at different sites, however, is the strong possibility that inhibitor concentrations at the sites are not equal in vivo. Why, for instance, should the concentration of a mercurial diuretic injected into an intact animal be identical at the tubular brush border and at the basolateral membrane?

Nevertheless, although the full toxicologic significance of differences in inhibitor activity at various sites may not always be clear, such activities frequently have been compared. One such instance is the mercury inhibition of tubular amino acid reabsorption in the kidney, as documented in Table 7-1. A reported difference between effects of inhibitors on basolateral and apical amino acid transport in the renal tubule is the inhibition of glutamate transport by succinate, observed only on the basolateral side in vivo (Foulkes, 1981). Similarly, only the uphill basolateral extrusion of Na in epithelial cells, mediated by Na,K-ATPase, is sensitive to inhibition by ouabain; downhill apical Na uptake is not so affected. On the other hand, the apical Na uptake is specifically depressed by amiloride.

A more compelling conclusion about inhibitor action at different sites may demand separation of the membranes in vitro (see Chapter 3) and individual study of their reactions with inhibitors. Such studies were performed, for instance, with phlorizin, an inhibitor of renal glucose transport. The drug inhibits the action of apical glucose transporters but also depresses basolateral uptake both in intact animals (Silverman, Aganon, and Chinard, 1970) and in purified vesicles (Kinne et al., 1975). Similarly, the active and Probenecid-sensitive basolateral accumulation of organic anions in the proximal tubule of mammalian kidneys may be contrasted with the much less Probenecid-sensitive and nonconcentrative transport of these anions across the apical membrane (Foulkes, 1963). In the intestine, the exit pathway of hexoses across the basolateral membrane differs from the active uptake at the apical side (Muekler, 1994). The mechanism of basolateral extrusion of heavy metals from enteric cells also differs from that of apical metal uptake (see section 6-3).

Numerous further instances have been reported in which waste products, xeno-biotics, or other solutes are exported from cells; selected examples are discussed in sections 7-2 and 7-3. Particular emphasis is placed on the role of the permeability gly-coprotein (Pgp) in the phenomenon of multidrug resistance. Induction or activation of a mechanism for extruding a particular drug or toxic agent from cells in response to exposure renders them more resistant to the agent; this essentially represents a compensatory or adaptive response of the cells.

Compensatory processes rapidly activated as a result of changes in the environment, or slower selective adaptation following more chronic exposure, frequently rest on al-terations in membrane function. Indeed, adaptive changes in membrane permeability and solute transport often have been reported. The phenomenon is summarized in sec-tion 7-4, with special emphasis on the potential of toxic interference with adaptation. The cellular response to altered environments is triggered by chemical or physical challenges and permits cells to develop, if necessary, appropriate new steady states under the widely varying conditions generally compatible with life. Toxic action of many agents can interfere with the attainment of such steady states and thus lead to permanent cell damage or death.

The cell volume represents one of these steady states, maintained over a range of osmotic insults by regulatory shrinkage or swelling of cells, as further described in Chapter 8. This focuses on the changes in membrane function responsible for volume regulation and on the action of many toxicants on this process. The association between cytotoxicity and swelling of cells is, of course, well known.

Membrane transport also has been found to facilitate export of selected metabolic waste products from cells. An instance of such an action is the excretion of lactic acid from working muscle cells, as discussed in section 7-2. Failure to remove lactic acid leads to intracellular lactate acidosis with consequent reduction of the efficiency of muscular work. Transfer of lactate across the cell membranes of rat diaphragm is me-diated by a relatively specific monocarboxylate carrier system sensitive to metabolic inhibition (Foulkes and Paine, 1961).

Solute extrusion from cells includes the process of exocytosis referred to in section 2-6 in relation to the effects of toxicants on membrane traffic. Exocytosis plays a role primarily in the export of macromolecules and is not discussed further here. One of the most extensively studied examples of extrusion of smaller ions and molecules is that of the carrier-mediated transport of sodium against its electrochemical activity gradient out of cells. This process is catalyzed by Na,K-ATPase, whose inhibition greatly affects cell composition and function. Similarly, inhibition of the membrane solute pump responsible for drug export and multidrug resistance (see section 7-3) damages cells by increasing their intracellular exposure to certain drugs. In both

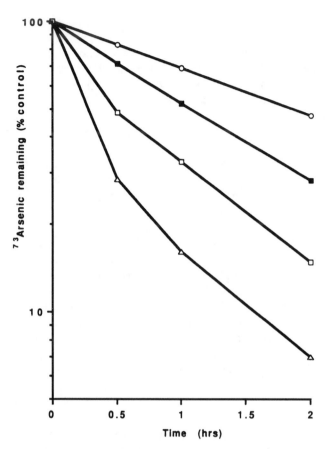

FIG. 7-1. Correlation between As extrusion and As resistance in Chinese hamster cells. Hyper-sensitive strain (*circles*), wild type control (*dark squares*), and two resistant strains (*light squares* and *triangles*) were preloaded with As and then incubated for 2 hours in As-free medium. (From Wang et al., 1996; with permission.)

instances, interference with solute extrusion can lead to significant cell damage. In summary, the inhibition of solute extrusion constitutes an important area in membrane toxicology.

7-2 TOXICOLOGIC SIGNIFICANCE OF MEDIATED
SOLUTE EXTRUSION

Cellular extrusion of xenobiotics reduces their intracellular concentration and thus is likely to increase cellular resistance to their toxic action. An example of such association of extrusion and resistance is illustrated in Fig. 7-1 by arsenic-resistant hamster cell lines. The resistant cells reduce intracellular concentrations of arsenic by an energy-requiring pump whose activity is inversely correlated with As sensitivity (Wang et al., 1996). The results shown were obtained with cells that first had been depleted of endogenous energy reserves to allow loading with arsenite. They next were washed extensively in the cold to remove extracellular arsenite, and finally they were incubated in arsenite-free media. Note that the hypersensitive strain (As/S5) retained arsenic most effectively, whereas the major portion of cellular As had been lost from the resistant cells (As/R27, As/R7) after 2 hours.

Similar examples of cells whose ability to export specific toxicants makes them insensitive to these agents are antibiotic-resistant bacteria, arsenic-resistant mutant species of *Leishmania* (Papadopoulou et al., 1994), and drug-resistant tumor cells (Dano, 1973). The extrusion is mediated by transporters, at times of admittedly broad specificity, and the transport systems themselves may become subject to inhibition by toxic agents. If such inhibition occurs, it results in increased cellular toxicant levels and consequently greater sensitivity to intoxication; the inhibitors are accordingly defined as chemosensitizing agents.

The sensitivity of export proteins like P glycoprotein (Pgp), and other resistance-associated transporters, to a variety of inhibitors leads to the possibility of increasing the sensitivity of cells to poisons or therapeutic drugs that normally are extruded. This effect possesses important clinical implications, especially in preventing the aquisition of resistance to antineoplastic drugs by cancer cells during chemotherapy and in the development of antibiotic-resistant microorganisms. Chemosensitizers that act by inhibiting Pgp action include Ca-channel blockers such as verapamil, quinidine, reserpine, cephalosporins, cyclosporin, and steroids (Gottesman and Pastan, 1993).

The extrusion of toxic solutes from cells frequently is catalyzed by a member of a specific drug-exporter family. Different such families have been described in various cells, but the most thoroughly explored is that of the so-called permeability glycoprotein discussed in the next section. Among others are the several carriers responsible for the export of tetracycline from bacterial cells (Yamaguchi et al., 1995). The substrate of these "tetracycline-resistance proteins" is a 1:1 complex of the drug with a divalent cation like Co^{++} or Cd^{++} (Yamaguchi, Udagawa, and Sawai, 1990). Presence of membrane-associated metal-efflux proteins can explain the lack of sensitivity of other bacterial strains to toxic metals like Cd (Silver and Walderhaug, 1992). A Cd-resistance protein in yeast has been described that also belongs to this family (Szczypka et al., 1994). Transmembrane transport of heavy metals in the form of specific complexes has been suggested repeatedly (see also Chapter 6), and formation of such complexes may provide an alternative mechanism for extruding such metals from cells.

Other members of related families of drug exporters include the cystic fibrosis transmembrane conductance regulator (CFTR) from humans, the human multidrug resistance–associated protein, and various other glycoproteins. Multispecific organic anion transporters have been described in several cell types (Saxena and Henderson, 1996); they are ATP-requiring pumps mediating the cellular extrusion of anionic detoxification products including glucuronates and mercapturates.

Solute extrusion from cells is important not only in the case of toxic xenobiotics but also for the normal export of waste products. An example of such handling of unwanted solutes is the excretion of lactate from working muscle cells. Lactate, the end product of muscle glycolysis during oxygen debt, must be removed from working muscle in order to prevent intracellular acidosis; accumulation of lactic acid in muscle cells greatly reduces the efficiency of contraction. Mediation of transmembrane movement of lactate and other anionic metabolites has been reported repeatedly. Foulkes and Paine (1961), for instance, found that the process of lactate transport by cells in the rat diaphragm is mediated by a monocarboxylate carrier possessing considerable structural specificity; it is sensitive (directly or indirectly) to inhibition by mercuric chloride, azide, and dinitrophenol.

Another instance in which cells normally export toxic or waste substances is the well-established contribution of intestinal pathways of elimination to systemic excretion of xenobiotics (Conradi, Burton, and Borchardt, 1996). Provided that this excretion follows transcellular rather than intercellular pathways, cell membranes presumably play a determining role in the process. A final example of mediated waste extrusion from cells is that described by Paulusma et al. (1996), who found in the rat that biliary excretion of a variety of anionic metabolites requires an ATP-dependent multispecific organic anion transporter (multidrug resistance–associated protein; see section 7-3) in the apical membrane of hepatocytes. Rats congenitally deficient in this protein are characterized by jaundice.

Cellular solute extrusion, further, represents the final step in solute transfer across cellular barriers like the renal tubular or intestinal epithelium or the capillary endothelium in brain (the blood–brain barrier; see section 6-7). The extrusion itself may be inhibitor sensitive, as in the instance of the Hg inhibition of renal cycloleucine reabsorption (see Table 7-1); similar results have been obtained with Cd (Foulkes and Blanck, 1991). Note again that we are concerned here with transcellular movement rather than diffusion between cells. For many, although not all, nonlipophilic solutes such extrusion results from mediation by membrane transport mechanisms. An exception appears to be the extrusion of heavy metals from enteric cells across the basolateral cell membrane during intestinal absorption (step 2 of absorption; see section 6-3). There is no evidence to suggest participation of any transport system in this process.

A classic example of mediated extrusion, on the other hand, is the basolateral export of Na during renal tubular reabsorption or intestinal absorption. This is a process catalyzed by Na,K-ATPase and specifically inhibited by ouabain. The apical extrusion of organic anions like PAH secreted into the tubular urine also appears to be a mediated process (Foulkes, 1977). Although both basolateral uptake and apical extrusion of PAH are sensitive to inhibition by Probenecid, the carrier systems involved are not identical.

A final example of inhibitor-sensitive mediated transport of a solute out of cells is the basolateral extrusion of amino acids during their reabsorption from the glomerular filtrate. Results summarized in Table 7-1 suggest that both apical uptake and basolateral extrusion of these amino acids are mediated steps sensitive to inhibition by Hg (Foulkes, 1987). Note that in this study Hg reduced the glomerular filtration rate by over two thirds and the fractional reabsorption of the filtered amino acid analogue cycloleucine from over 99% to 74%. Absolute reabsorption of cycloleucine (the reabsorptive flux) decreased from 100 to 22 nmol/min, while its retention in the tissue rose from 74 to 210 nmol; in other words, cycloleucine was trapped in the cells. As a result, the observed time required for the amino acid to cross the cell from tubular lumen into renal venous blood (the reabsorption delay) was lengthened greatly, from 40 to more than 120 seconds. Note that the observed delays agree well with those predicted on the basis of the linear model of transepithelial movement (see Fig. 3-2) and of empirical values for the tissue accumulation and the rate of amino acid reabsorption. These findings support the conclusion that Hg inhibited the removal of cycloleucine from tubular urine at the brush border (reabsorptive flux), as well as its subsequent basolateral extrusion.

As mentioned, the ability to extrude xenobiotics from cells is often induced in response to toxic exposures and therefore constitutes a specific example of the wider phenomenon of cell adaptation to changes in the environment. This does not imply that induction of membrane transport systems or alterations in membrane function always must be involved in cell adaptation to toxic agents, a fact further emphasized in section 7-4. The conclusion, however, that membrane functions participating in the process of cellular adaptation may be themselves sensitive to toxic effects offers yet another potential mechanism of cytotoxic action at the cell membrane.

7-3 MULTIDRUG RESISTANCE AND P GLYCOPROTEIN

Juliano and Ling (1976) first characterized a surface glycoprotein in Chinese hamster ovary cells whose presence in various mutants was correlated with resistance to certain drugs. Because of its apparent role in the control of cell permeability to drugs, the protein was named P glycoprotein. Its molecular weight was estimated at 170,000 D; the biochemical properties of Pgp are discussed in detail in the review of Gottesman and Pastan (1993). Association of drug resistance with activity of the P glycoprotein is a fairly general phenomenon which has been observed in a variety of cells (see section 1). The observation that the protein is present in membrane vesicles prepared from multidrug-resistant cells (Lelong et al., 1992) strengthens the view that Pgp plays an important role in membrane transport of xenobiotics.

Not all examples of multidrug resistance, however, can be attributed to the activity of P glycoprotein. For instance, Cole et al. (1992) worked with a multidrug-resistant human cancer cell line and observed overexpression of a gene coded for a transporter protein different from Pgp. Like Pgp, it possesses a high molecular weight (1522 amino acid residues). The two proteins differ significantly in their structures, however, and exhibit different substrate specificities. Nor is the new protein affected by cyclosporin or some of the other chemosensitizing agents known to inhibit transport of xenobiotics by Pgp. Chemosensitizers, as already mentioned, derive their name from their ability to increase the sensitivity of cells to toxic xenobiotics by inhibiting an extrusion pump. The multidrug-resistance protein described by Slapac et al. (1996) and expressed in plasma membranes of murine cells also differs from Pgp.

Generally, however, the multidrug-resistant phenotype is characterized by overexpression of Pgp, relatively broad drug resistance, and increased drug sensitivity in presence of Pgp inhibitors like verapamil and cyclosporin. For instance, Ernest and Bello-Reuss (1995) observed that the ED_{50} of adriamycin in a mouse kidney cell line approximates 1 μM in controls but is reduced to about 0.2 μM in presence of cyclosporin (10 μM). The explanation offered for these findings implicitly assumes that the two toxicants do not exert independent and additive effects.

Pgp also influences movement of xenobiotics across cell barriers like the blood–brain barrier. For instance, Schinkel et al. (1994) reported that mice in which the gene for Pgp had been deleted were approximately 100-fold more sensitive to the neurotoxic agent ivermectin than was the original strain. This finding suggests that ivermectin, after penetrating cells of the capillary endothelium, normally largely is returned to plasma by Pgp. Brain capillary endothelial cells indeed have been reported to contain significant amounts of Pgp at their luminal membranes (Cordon-Cardo et al., 1989).

A similar role of Pgp has been invoked to explain the exclusion of cephalosporin from the central nervous system (Tsuji et al., 1993). Although this drug is highly lipid-soluble and therefore should be readily able passively to cross cell membranes, it actually does not diffuse normally across the blood–brain barrier. An explanation for this finding is suggested by the further observation that net extraction of cephalosporin from arterial blood in rat brain is greatly increased when ATP levels required for Pgp activity are reduced by transient ischemia, or in the presence of the Pgp inhibitor quinidine (Sakata, Tamai, and Kawazu, 1994). Normally the concentration of cephalosporin in the barrier cells is presumably kept low by its P glycoprotein–mediated extrusion across the apical membrane; the increased cerebral retention of cephalosporin following depression of Pgp activity results from the reduced return of the drug from cells to blood. The consequent rise in the intracellular concentration of the drug apparently further favors its basolateral leakage into the underlying tissue.

Surprisingly, as observed by Stein et al. (1994), influx and efflux of drugs can be affected differentially by P glycoprotein; this is illustrated in Fig. 7-2. Here a colchicine-sensitive parent strain of mouse cells (strain 1) was transfected with vectors encoding wild-type (strain 2) or mutant (strain 3) multidrug transporter. Strain 3 proved more resistant to colchicine than strain 2 and exhibited a slower initial rate of colchicine uptake.

The model of Pgp action suggested by Stein et al. (1994) for the different effects of Pgp on drug influx and efflux visualizes the transport protein extracting drugs directly from the two leaflets of the membrane bilayer; the outer leaflet is assumed to be in rapid equilibrium with the extracellular fluid, and the inner leaflet similarly equilibrates with cytoplasm. Pgp is thought to reduce the initial rate of drug uptake by extruding it from the outer layer. Such an action would depend on a relatively high binding affinity of Pgp for its substrates and inhibitors, as indeed reported by Gottesman and Pastan (1993) and Rao and Scarborough (1994). It also is assumed implicitly by this model that the drugs undergoing bidirectional transport possess a finite solubility in the membrane. In any case, multidrug-resistant cells appear to be

(A) COLCHICINE

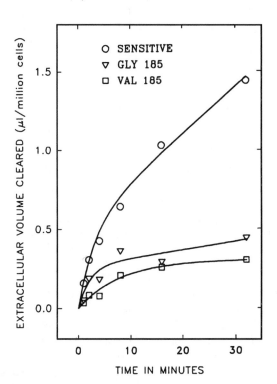

FIG. 7-2. P glycoprotein–associated colchicine uptake. The time course of colchicine uptake is shown in a sensitive parent strain of mouse cells and two transfected strains, wild-type (*triangles*) and mutant (*squares*). (From Stein et al., 1994; with permission.)

able to maintain low intracellular concentrations of certain drugs both as a result of increased drug extrusion and of decreased drug influx.

A review by Higgins (1993) summarizes many of the functions of Pgp. The protein is thought to act essentially as an energy-requiring pump able to extrude a variety of xenobiotics from the cell. The hypothesis of such a pump activity is widely supported by many observations. For instance, Jancis et al. (1993) observed that the production of mRNA for Pgp is stimulated on exposure of rat pituitary tumor cells to estradiol. This permits control cells to retain five to ten times more of the fluorescent Pgp substrate rhodamine than do estrogen-treated cells. The difference disappears in presence of the Pgp inhibitors verapamil (1 μM) or cyclosporin (0.3 μM). Figure 7-3 provides some details of these results.

The concept of Pgp acting as drug pump, requiring ATP to catalyze the active extrusion of drugs against their activity gradient, raises some unanswered questions, however. One conceptual difficulty arises from the great variety of drugs involved. The Pgp substrates generally are preferentially soluble in lipids and include perhaps

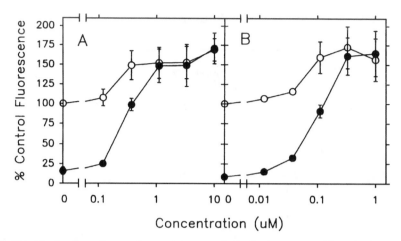

Concentration (uM)

FIG. 7-3. Rhodamine extrusion from rat tumor cells by P glycoprotein. The abscissa shows concentrations of verapamil in (a) and cyclosporin in (b). Control cells are *open circles*, estradiol-treated cells are *closed circles*. (From Jancis et al., 1993; with permission.)

hundreds of natural products and synthetic organic compounds (Gottesman and Paston, 1993). Such broad specificity, as pointed out by Roepe (1995), certainly exceeds that of typical carrier–substrate interactions. Alternative models for the action of Pgp have been proposed (see Hoffman, Wei, and Roepe, 1996).

Glutathione S-conjugates also are transported by the multidrug resistance–associated protein (Jedlitschky, Leier, and Buchholz, 1994), a finding compatible with the observation of Huang and Lee (1996) that the thiol compound is required for arsenite extrusion. This extrusion is depressed by verapamil and other inhibitors of Pgp activity. It must be pointed out, however, that the observed involvement of glutathione could be explained alternatively by postulating an indirect role of the thiol compound. This could simply answer the need to maintain arsenite in a diffusible form, not bound to cytoplasmic protein. In that case other small ligands should be as effective as glutathione. Their role would be similar to that of, for example, mercaptoethanol required for renal uptake of heavy metals injected into the renal artery (Foulkes, 1974); in absence of a suitable low–molecular weight ligand, the injected metals are instantaneously bound to plasma proteins and become unavailable to the kidney.

The ATP requirement of P glycoprotein–mediated extrusion of xenobiotics makes the process sensitive to metabolic inhibitors. A major role of ATP here is to maintain the solute channel in an open form. A number of channels are similarly sensitive to ATP (see, e.g., Mayorga-Wark, Dubinsky, and Schultz, 1995). Bosch et al. (1996) found that the expression of *Drosophila melanogaster* P glycoproteins is associated with ATP-channel activity. Other evidence also indicates a function of Pgp as ATP channel (Abraham et al., 1993). The physiologic significance of a channel-mediating loss of ATP from cells may be related to functions of extracellular ATP and the presence of purine receptors on cells.

This view is strengthened by the finding that nonhydrolyzable ATP analogs in several instances can substitute for ATP in supporting activity of different membrane channels. Although the role of ATP in these cases, therefore, does not involve hydrolysis or phosphorylation, Pgp can act as drug-stimulated membrane ATPase (Sarkadi et al., 1992). Ambudkar et al. (1992) described the characteristics of drug-stimulated ATP hydrolysis by a partially purified and reconstituted human multidrug-resistance pump.

In addition to its capacity for xenobiotic transport, the glycoprotein exhibits properties of a transmembrane chloride channel (Higgins, 1993). These two functions, however, seem distinct and separate. The association of Pgp with chloride channel activity, as is the case with the related cystic fibrosis transmembrane conductance regulator, points to a possible involvement of Pgp in cell-volume regulation (Valverde et al., 1994). Cell-volume regulation and its inhibition by toxic agents are discussed in Chapter 8.

A further possible function of Pgp has been suggested by the work of Charuk et al. (1994) on renal secretion of drugs into urine. This may reflect active extrusion of drugs by Pgp across the apical cell membrane into the tubular lumen. Transepithelial drug-transport pathways apparently involving Pgp have been demonstrated in monolayers of renal cell lines (e.g., Horio et al., 1990). The fact that renal secretion of digoxin in vivo is inhibited by cyclosporin and quinidine (De Lannoy et al., 1992) also may reflect a contribution of P glycoprotein–mediated transport, but the evidence for involvement of Pgp in this process remains incomplete.

A possible general function of Pgp in secretion, however, is suggested by the observation that the protein is highly expressed in other secretory organs such as the adrenals and colonic epithelium (Cole et al., 1992). Secretion of a drug in this manner becomes an important determinant of its overall toxicokinetic behavior, a further example of the general role of membranes in toxicokinetics (see also section 6-2).

7-4 INDUCIBLE SOLUTE TRANSPORT AND CELL ADAPTATION

To the extent that development of multiple drug resistance is associated with an increase in the concentration of a specific membrane-carrier system like Pgp, the process illustrates how alterations in membrane function permit cells to compensate for, or adapt to, changes in their environment. The onset of compensatory activity usually is rapid, as in the instances described in section 4-3 and in Chapter 8; adaptation, in contrast, is a slower process that may become overt only after a significant initiation delay.

Examples of membrane responses to osmotic stress are discussed in the next chapter. The specific regulation of intestinal nutrient transporters by their dietary substrates is reviewed by Ferraris and Diamond (1989) and is referred to in section 4-3. Transport capacity for certain nutrients may decrease as a result of their absence from an otherwise adequate diet. In contrast, the efficiency with which a nutrient can be taken up by cells may be increased during starvation; this condition stimulates, for instance, the synthesis of neutral amino acid carriers (system A) in the plasma membranes of mammalian cells. This adaptive up-regulation, as described by Kilberg, Hutson, and Lane (1994), is depressed by inhibitors of protein synthesis, including especially that of glycoproteins, as well as of RNA metabolism.

Unsaturated fatty acids in the diet may influence function by their direct incorporation into the membrane. Such a change in membrane composition is likely to affect membrane fluidity, an important presumptive factor in absorption, for instance, of heavy metals (see sections 6-3 and 6-8). Because fluidity varies significantly and directly with temperature, an important role of fluidity in a transport process implies a likelihood that the process will prove to be temperature dependent. Accordingly, as previously emphasized, it would not be appropriate to identify the temperature-dependence of a membrane process, that is, a Q_{10} value greater than 1.0, as adequate proof that it is a mediated, let alone an active, process directly tied to a source of metabolic energy. Such an identification has been made at times, as mentioned in section 6-2, for instance in studies of mercury uptake by kidney cells (Endo, Sakata, and Shaikh, 1995) and of Cd accumulation by human liver cells (Souza, Bucio, and Gutierrez-Ruiz, 1997). This is not, however, a satisfactory interpretation of the temperature-dependence of solute transfer.

Physiologic adaptation to new environments also may be achieved by endocrine responses. To the extent that hormones regulate cell function by controlling membrane solute transport, this provides another role for membranes in the response to changes in the environment. Stimulation of transport at the cell membrane may involve insertion of existing or newly synthesized transporters from within the cell. If adaptation includes induction of protein synthesis, inhibitors like oligomycin are likely to interfere with the adaptive process.

Adaptation and compensation may be observed both at the level of intact organisms and also with single cells. The adaptive response of cells to the presence of toxic agents in their environment is illustrated in section 7-3 by the acquisition of P glycoprotein–related drug resistance. In that case verapamil and other Pgp inhibitors depress adaptation.

None of these facts, of course, imply a necessary role of cell membranes in all instances of cellular responses to a new environment. A well-documented example of a response not based on membrane reactions is the intracellular induction of metallothionein synthesis triggered by exposure to heavy metals; metallothionein concentrations frequently have been correlated with metal tolerance. Metallothionein is primarily a cytoplasmic protein, so that the adaptive rise in its cellular concentration is not tied to changes in membrane function.

Altering the net rate of penetration of a xenobiotic across the cell membrane, or the extent to which it is retained in the cells, determines its concentration at intracellular target sites. To the many membrane-related variables that can modulate the toxicity of a xenobiotic, such as rates of absorption, clearance, cellular uptake, and extrusion, must be added possible compensatory or adaptive changes at the cell membrane. All of these are potentially important factors that can determine risks attending exposure to a toxic compound.

REFERENCES

Abraham, E.H., A.G. Prat, L. Gerweck, T. Seneveratne, R.J. Arceci, R. Kramer, G. Guidotti, and H.F. Cantiello. 1993. The multidrug resistance (mdr1) gene product functions as an ATP channel. *Proc. Natl. Acad. Sci. U.S.A.* 90:312–316.

Ambudkar, S.V., I.H. Lelong, J. Zhang, C.O. Cardarelli, M.M. Gottesman, and I. Pastan. 1992. *Proc. Natl. Acad. Sci. U.S.A.* 89:8472–8476.

Bosch, I., G.R. Jackson, J.M. Croop, and H.F. Cantiello. 1996. Expression of Drosophila melanogaster P-glycoproteins is associated with ATP-channel activity. *Am. J. Physiol.* 271:C1527–C1538.

Charuk, J.H.M., T.W. Loo, D.M. Clarke, and R.A.F. Reithmeier. 1994. Interaction of rat kidney P-glycoprotein with a urinary component and various drugs including cyclosporin A. *Am. J. Physiol.* 266:F66–F75.

Cole, S.P.C., G. Bhardwaj, J.H. Gerlach, J.E. Mackie, C.E. Grant, K.C. Almquist, A.J. Stewart, E.U. Kurz, A.M.V. Duncan, and R.G. Deeley. 1992. Overexpression of a transporter gene in a multidrug resistant human lung cancer cell line. *Science* 258:1650–1654.

Conradi, R.C., P.S. Burton, and R.T. Borchardt. 1996. Physico-chemical and biological factors that influence a drug's cellular permeability by passive diffusion. In *Lipophilicity in drug action and toxicology*, eds. V. Pliska, B. Testa, and H. van de Waterbeemd, 232–253. New York: VCH Publishers.

Cordon-Cardo, C., J.P. O'Brien, D. Casals, L. Rittman-Grauer, J.L. Biedler, M.R. Melamed, and J.R. Bertino. 1989. Multidrug-resistance gene (P-glycoprotein) as expressed by endothelial cells at blood–brain barrier sites. *Proc. Natl. Acad. Sci. U.S.A.* 86:695–698.

Dano, K. 1973. Active outward transport of daunomycin in resistant Ehrlich ascites tumor cells. *Biochim. Biophys. Acta* 323:466–483.

De Lannoy, I.A.M., G. Koren, J. Klein, J. Charuk, and M. Silverman. 1992. Cyclosporin and quinidine inhibition of renal digoxin excretion: Evidence for luminal secretion of digoxin. *Am. J. Physiol.* 263:F613–F622.

Endo, T., M. Sakata, and Z.A. Shaikh, 1995. Mercury uptake by primary cultures of rat renal cortical epithelial cells: I. Effects of cell density, temperature and metabolic inhibitors. *Toxicol. Appl. Pharmacol.* 132:36–43.

Ernest, S., and E. Bello-Reuss. 1995. Expression and function of P-glycoprotein in a mouse kidney cell line. *Am. J. Physiol.* 269:C323–C333.

Ferraris, R.P., and J.P. Diamond. 1989. Specific regulation of intestinal nutrient transporters by their dietary substrates. *Ann. Rev. Physiol.* 51:125–141.

Foulkes, E.C. 1963. The kinetics of PAH secretion in the rabbit. *Am. J. Physiol.* 205:1019–1024.

Foulkes, E.C. 1974. Excretion and retention of cadmium, zinc and mercury by the rabbit kidney. *Am. J. Physiol.* 227:1356–1360.

Foulkes, E.C. 1977. Movement of PAH between lumen and cells of renal tubule. *Am. J. Physiol.* 232: F424–F428.

Foulkes, E.C. 1981. Asymmetry of membrane functions in transporting cells. In *Renal transport of organic substances*, eds. R. Greger, F. Lang, and S. Silbernagl, 45–54. Berlin: Springer Verlag.

Foulkes, E.C. 1987. Role of basolateral membranes in organic solute reabsorption in rabbit kidney. *Am. J. Physiol.* 252:F1042–F1047.

Foulkes, E.C., and S. Blanck. 1991. Cadmium inhibition of basolateral solute fluxes in rabbit renal tubules and the nature of cycloleucine uptake. *Toxicol. Appl. Pharmacol.* 108:150–156.

Foulkes, E.C., and C.M. Paine. 1961. The uptake of monocarboxylic acids by rat diaphragm. *J. Biol. Chem.* 236:1019–1022.

Gottesman, M.M., and I. Pastan. 1993. Biochemistry of multidrug resistance mediated by the multidrug transporter. *Ann. Rev. Biochem.* 62:385–427.

Higgins, C.H. 1993. The multidrug resistance P-glycoprotein. *Curr. Opinion Cell Biol.* 5:684–687.

Hoffman, M.M., L.Y. Wei, and P.D. Roepe. 1996. Are altered pH and membrane potential in huMDR1 transfectants sufficient to cause MDR protein-mediated multidrug resistance? *J. Gen. Physiol.* 108:295–313.

Horio, M., I. Pastan, M.M. Gottesman, and J.S. Handler. 1990. Transepithelial transport of vinblastine by kidney-derived cell lines. *Biochim. Biophys. Acta* 1027:116–122.

Huang, R.N., and T.C. Lee. 1996. Arsenite efflux is inhibited by verapamil, cyclosporin A, and GSH-depleting agents in arsenite-resistant Chinese hamster ovary cells. *Toxicol. Appl. Pharmacol.* 141:17–22.

Jancis, E.M., R. Carbone, K.J. Loechner, and P.S. Dannies. 1993. Estradiol induction of rhodamine 123 efflux and the multidrug resistance pump in rat pituitary tumor cells. *Mol. Pharmacol.* 43:51–56.

Jedlitschky, G., I. Leier, U. Buchholz, M. Center, and D. Keppler. 1994. ATP-dependent transport of glutathione S-conjugates by the multidrug resistance-associated protein. *Cancer Res.* 54:4833–4836.

Juliano, R.L., and V. Ling. 1976. A surface glycoprotein modulating drug permeability in Chinese hamster ovary cell mutants. *Biochim. Biophys. Acta* 455:152–162.

Kilberg, M.S., R.G. Hutson, and R.O. Laine, 1994. Amino acid-regulated gene expression in eucaryotic cells. *FASEB J.* 8:13–19.

Kinne, R.K.H., H. Murer, E. Kinne-Saffran, M. Thees, and G. Sachs. 1975. Sugar transport by renal plasma membrane vesicles. *J. Membr. Biol.* 21:375–395.

Lelong, I.H., R. Padmanabhan, E. Lovelace, I. Pastan, and M.M. Gottesman. 1992. ATP and GTP as alternative energy sources for vinblastine transport by P-170 in KB-V1 plasma membrane vesicles. *FEBS Lett.* 304:256–260.

Mayorga-Wark, O., W.P. Dubinsky, and S.G. Schultz. 1995. Reconstruction of a K_{ATP} channel from basolateral membranes of Necturus enterocytes. *Am. J. Physiol.* 269:C464–C471.

Muekler, M. 1994. Facilitative glucose transporters. *Eur. J. Biochem.* 219:713–725.

Papadopoulou, B., G. Roy, S. Dey, B.P. Rosen, and M. Ouellette. 1994. Contribution of the Leishmania P-glycoprotein-related gene ltpgpA to oxyanion resistance. *J. Biol. Chem.* 269:11980–11986.

Paulusma, C.C., P.J. Bosma, G.J.R. Zaman, C.T.M. Bakker, M. Otter, G.L. Scheffer, R.J. Scheper, P. Borst, and R.P.J. Oude Elferink, 1996. Congenital jaundice in rats with a mutation in a multidrug resistance-associated protein gene. *Science* 271:1126–1128.

Rao, U.S., and G.A. Scarborough. 1994. Direct demonstration of high affinity interactions of immunmosuppressant drugs with the drug binding site of the human P-glycoprotein. *Mol. Pharmacol.* 45:773–776.

Roepe, P.D. 1995. Arguments against the drug-pump model for multidrug resistance. *J. NIH Res.* 7:14–15.

Sakata, A., I. Tamai, K. Kawazu, Y. Deguchi, T. Ohnishi, A. Saheki, and A. Tsuji. 1994. In vivo evidence for ATP-dependent and P-glycoprotein-mediated transport of cyclosporin A at the blood-brain barrier. *Biochem. Pharmacol.* 48:1989–1992.

Sarkadi, B., E.M. Price, R.C. Boucher, U.A. Germann, and G.A. Scarborough. 1992. Expression of the human multidrug resistance cDNA in insect cells generates a high activity drug-stimulated membrane ATPase. *J. Biol. Chem.* 267:4854–4858.

Saxena, M., and G.B. Henderson. 1996. MOAT4, a novel multispecific anion transporter for glucuronides and mercapturates in mouse L1210 cells and human erythrocytes. *Biochem. J.* 320:273–281.

Schinkel, A.H., J.J. Smit, O. van Tellingen, J.H. Beijnen, E. Wagenaar, L. van Deemter, C.A. Mol, M.A. van der Valk, E.C. Robanus-Maandag, H.P.J. te Riele, A.J.M. Berns, and P. Borst. 1994. Disruption of the mouse mdr1a P-glycoprotein gene leads to a deficiency in the blood-brain barrier and to increased sensitivity to drugs. *Cell* 77:491–502.

Silver, S., and M. Walderhaug, 1992. Gene regulation of plasmid- and chromosome-determined inorganic ion transport in bacteria. *Microbiol. Rev.* 56:195–228.

Silverman, M., M.A. Aganon, and F.P. Chinard. 1970. D-glucose interaction with renal tubule cells. *Am. J. Physiol.* 218:735–742.

Slapac, C.A., R.L. Martell, M, Terashima, and S.B. Levy. 1996. Increased efflux of vincristine, but not of daunorubicin, associated with the murine multidrug resistance protein. *Biochem. Pharmacol.* 52:1569–1576.

Souza, V., L. Bucio, and M.C. Gutierrez-Ruiz. 1977. Cadmium uptake by a human hepatic cell line. *Toxicology* 120:215–220.

Stein, W.D., C. Cardarelli, I. Pastan, and M.M. Gottesman. 1994. Kinetic evidence suggesting that the multidrug transporter differentially handles influx and efflux of its substrates. *Mol. Pharmacol.* 45:763–772.

Szczypka, M.S., J.A. Wemmie, W.S. Moye-Rowley, and D.J. Thiele. 1994. A yeast metal resistance protein similar to human cystic fibrosis transmembrane conductance regulator (CFTR) and multidrug resistance-associated protein. *J. Biol. Chem.* 269:22853–22857.

Tsuji, A., I. Tamai, A. Sakata, Y. Tenda, and T. Terasaki. 1993. Restricted transport of cyclosporin A across the blood-brain barrier by a multidrug transporter P-glycoprotein. *Biochem. Pharmacol.* 46:1096–1099.

Valverde, M.A., M. Diaz, F.V. Sepulveda, D.R. Gill, S.C. Hyde, and C.F. Higgins, 1994. Volume-regulated chloride channels with the human multidrug-resistance P-glycoprotein. *Nature* 355:830–833.

Wang, Z., S. Dey, B.P. Rosen, and T.G. Rossman. 1996. Efflux-mediated resistance to arsenicals in arsenic-resistant and -hypersensitive Chinese hamster cells. *Toxicol. Appl. Pharmacol.* 137:112–119.

Yamaguchi, A., Y. Shiina, E. Fujihara, T. Sawai, N. Noguchi, and M. Sasatsu. 1995. The tetracycline efflux protein encoded by the (tet)K gene from St. aureus is a metal-tetracycline/H^+ antiporter. *FEBS Letters* 365:193–197.

Yamaguchi, A., T. Udagawa, and T. Sawai, 1990. Transport of divalent cations with tetracycline as mediated by the transposon Tn10-encoded tetracycline resistance protein. *J. Biol. Chem.* 265:4809–4813.

8

Cell Membranes in Volume Homeostasis and Effects of Toxicants

8-1 INTRODUCTION

Cell-volume homeostasis describes the general ability of most cells to maintain a relatively constant volume, even in the face of significant changes in the osmolarity of the extracellular fluid. An exception are mammalian nonnucleated erythrocytes, which over a considerable range of imposed osmolarities (from approximately 220 to 380 mosM) behave as perfect osmometers, their volume changing inversely with extracellular osmolarity. The present chapter discusses the critical part played by the plasma membrane in cells other than mammalian erythrocytes in the maintenance of steady cell volumes. The important role of membranes in volume control is seen clearly in the association of functional membrane lesions with cell swelling. Occurrence of cell swelling in poisoned or anoxic tissues is, of course, well known.

The emphasis placed here on membranes does not imply that cellular-volume homeostasis is governed exclusively by membrane-associated processes. Thus, extensive evidence shows that metabolism of organic osmolytes, as described in section 8-4, and factors associated with the cytoskeleton and the extracellular matrix also help determine cell volumes. A role of the cytoskeleton in toxicant-induced cell swelling is suggested, for instance, by the report that the cytotoxic metal cadmium both reacts with the cytoarchitecture (see section 5-6) and causes cell swelling. Such effects, however, are not considered further here except to the extent that they involve membrane function.

The plasma membrane surrounding the cytoplasm possesses little mechanical strength. As a result, in animal cells without the rigid support of an outer cell wall, variations in turgor pressure lead to immediate changes in cell volume. Moreover, the presence of aqueous pores in most cell membranes readily permits movement of water and thereby tends to abolish osmotic pressure differences between cells and extracellular fluid; this results in the tendency of cells in hypotonic media to swell. The assumption is made here that no pumps exist specifically to transport water against pressure or concentration gradients. Animal cells passively responding to hypertonic media, therefore, are expected to shrink; conversely, in hypotonic media the cells swell. The fact that most cells do not permit their volume to change appreciably under such conditions reflects the action of compensatory mechanisms other than water pumping to counteract the effects of an osmotic gradient across the membrane.

Historically, two major possible mechanisms were considered for how intracellular turgor pressure and consequently the volume of mammalian cells is maintained constant. One mechanism assumed a normally hypertonic cytoplasm, with continuous active extrusion of water. Alternatively, cells were considered to maintain their volume by controlling the osmolarity of the cytoplasm; in this case, homeostasis consists of regulation of solutes rather than water in the cell. These possibilities are further discussed in section 8-3. Both hypotheses assign a major role in volume homeostasis to function at the cell membrane and visualize the association of cell swelling and cytotoxicity in terms of inhibition of such function.

There is actually little evidence for active water transport (see section 8-3). Consequently, the osmolar composition of the cell generally is regarded as the major determinant of cell volume. Cytoplasmic osmolarity, in turn, is regulated by two major processes: One is the metabolic formation or breakdown of specific organic osmolytes (see section 8-4); the other consists of the control over loss or gain of solutes across the cell membrane. To the extent that cytoplasmic osmolarity in this manner results from the flow of solutes into and out of cells, cell volume therefore is a function of the net permeability of the membrane to a variety of solutes.

The need to emphasize net permeability can be illustrated by the fact that increased sodium flux into the cell would have no osmotic consequence in presence of stimulation of active sodium extrusion. However, any toxic agent directly or indirectly altering the net permeability of the membrane to intracellular or extracellular solutes thereby also can interfere with volume homeostasis. The same influence is seen if cold or anoxia or general metabolic inhibitors depress energy metabolism and consequently active solute transport at the membrane. The result, again, is the inability of the cell to regulate its volume, so that it swells even in isotonic solutions; section 8-3 discusses the mechanism of such isotonic swelling. It is important to remember in this connection how frequently cytotoxicity is accompanied by cellular edema; swelling of cells and subcellular organelles like mitochondria routinely are observed following toxic exposures. The critical role of membranes in volume homeostasis makes effects of toxic agents on cell volumes an important area in membrane toxicology (see section 8-6).

As pointed out, the absence of a supporting cell wall does not permit plasma membranes to withstand significant hydrostatic pressure differences. Nevertheless, animal cells generally do not passively shrink or swell in response to a broad range of imposed osmotic gradients; instead, they compensate by changes both in osmolyte metabolism (see section 8-4) and in membrane solute fluxes. The ability of membranes to control solute fluxes is thus critical both for the maintenance of steady-state cell volumes and for compensatory volume changes following exposure to an osmotic insult. The objective of compensatory activity is to shrink cells acutely swollen by exposure to hypotonic solutions, or to expand the volume of cells shrunk in hypertonic media. Like normal volume control, compensatory volume regulation is also sensitive to the actions of many toxic agents (see section 8-6).

Volume control and compensatory responses are not restricted to whole cells. Indeed, mitochondria also maintain constant volumes and react to changes in the osmolarity of their surroundings by appropriate compensatory actions. As in whole cells, the steady-state volume of these organelles is determined largely by the ability of the membranes to regulate solute influx and efflux, and interference with normal energy metabolism leads to swelling. Even though cell edema often is associated with swollen organelles, however, cellular and mitochondrial volume changes appear to represent independent processes that are not always coupled (see section 5-7).

No detailed review need be offered here of the field of cell-volume homeostasis, as this has been discussed frequently. Reference may be made, for instance, to the monograph of Lang and Haussinger (1993). A number of other informative reviews have been written on specific aspects of cell-volume control and its regulatory adjustments (e.g., Hoffmann and Simonsen, 1989). The present chapter briefly summarizes what is known of the nature of volume control by cells and subcellular organelles, and especially of the role played in it by membranes. It then deals with technical problems encountered in measuring cell edema and finally focuses on different mechanisms whereby toxic agents disrupt volume homeostasis.

8-2 DETERMINATION OF INTRACELLULAR VOLUMES

Detailed analysis of mechanisms responsible for volume changes requires an ability to measure intracellular solute concentrations and volumes of distribution. This becomes clear in considering the fact that swelling of cells usually, although not always (see section 5-7), involves not only the cytoplasm; indeed, swollen mitochondria frequently are associated with cytotoxicity. However even though a change in, for example, the volume of cytoplasm, may be a sensitive indicator of cell malfunction it is difficult in the living cell to determine the size of each of the fluid compartments (e.g., cytoplasmic, nuclear, mitochondrial) that make up the total intracellular water volume.

Commonly, cell swelling is measured by assay of total tissue water (in mL/g dry weight), corrected for adhering extracellular water by determination of, for instance, the inulin space. This correction, however, cannot define the intracellular fluid volume in all tissues. Total tissue water in the kidney includes tubular urine containing high concentrations of inulin in vivo, and perhaps zero inulin if the marker is added in vitro (see section 3-3). Estimation of the interstitial volume of distribution of an extracellular marker in vivo also is rendered difficult if a diffusion barrier like the blood–brain barrier limits uptake of inulin or other strictly extracellular markers.

The detailed analysis of cell swelling, in presence or absence of toxicants, is complicated further by the fact already referred to that knowledge of total cell water provides little information on the contribution of cytoplasm and subcellular organelles to the intracellular water volume. Without information on cytoplasmic volume and the distribution of a solute between organelles and cytoplasm, the concentration of this solute in cytoplasm cannot be readily calculated. This fact is shown in section 3-3 to interfere, for example, with the estimation of the intracellular concentration achieved by organic anions actively transported by renal cells. In reference to volume homeostasis, the problem becomes important because, as described in section 5-7 in relation to the effects of toxicants on mitochondrial and other subcellular membranes, volume of cytoplasm and organelles may vary independently. A cytoplasmic volume probe like 3-O-methylglucose, which has been applied to renal cortex in vivo (Foulkes and Blanck, 1993) as well as to other tissues (Foulkes and Blanck, 1994), may prove useful in further explorations of subcellular volumes of distribution (see Chapter 5).

Additional uncertainty about cytoplasmic concentrations is introduced if an intracellular solute is sequestered in vesicles, such as those that may be involved in the transepithelial movement of organic anions in the kidney (Miller, Stewart, and Pritchard, 1993). As a result of all these problems, it is difficult to estimate the cytoplasmic concentrations of specific osmolytes; a similar difficulty is discussed (see section 5-7) in reference to estimating inhibitor concentrations at intracellular target sites such as mitochondrial membranes in the living cell.

8-3 THE PROBLEM

Before discussing how toxic agents can affect cell volume, it is useful to recapitulate the basic mechanisms governing bulk water movement in and out of cells. The question of the osmolarity of cell contents remained a very controversial topic for many years. Although a tissue slice suspended at 37°C in oxygenated Ringer's solution can maintain its normal water content of around 75% (v/w), anoxia or cooling of the preparation causes the tissue to swell. Theoretically this could be explained by assuming a normal hypertonicity of cell contents, maintained by continuous active extrusion of water. Alternatively, as now generally accepted, cell contents are normally isotonic with extracellular medium. In this case, the metabolism-dependent and inhibitor-sensitive maintenance of intracellular osmolarity and therefore cell volume reflects the constant need actively to extrude primarily the sodium ion, which passively enters the cell along its electrochemical activity gradient.

These hypotheses became directly testable only when methods were developed to measure the osmolarity of cell contents in the absence of artifacts arising, for example, from the hydrolysis of labile compounds like creatine phosphate in muscle and other rapid chemical changes initiated on destruction of cells. In one such technique freshly collected tissue is rapidly plunged into a known volume of boiling water; denatured protein is spun off and the osmolarity of the supernatant determined. The results obtained by this and similar techniques generally support the conclusion that mammalian tissues like liver and muscle are isotonic with plasma.

The following plausible explanation for cell swelling in the absence of metabolism was suggested on that basis (Leaf, 1959): Suppression of active sodium extrusion permits its net inflow into cells; the resulting cell depolarization increases net influx of chloride; and uptake of sodium and chloride is accompanied by an osmotically equivalent volume of water. In other words, cells normally swell by uptake of isotonic saline. In the absence of normal Na levels, however, cells may swell if depolarization at the cell membrane permits influx of chloride ions; this occurs, for instance, in high-potassium Ringer's solution.

As pointed out, the role of membranes in volume regulation depends primarily on their contribution to the control of solute loss or gain by cells. Quite generally, therefore, inhibition of active solute transport at the membrane, even in isotonic solutions, leads to cell swelling. It is important here to reiterate that, in studying such effects, care must be taken not to conclude either that temperature-dependence constitutes proof for active transport (see section 6-2) or that a solute is being actively transported solely on the basis that its movement is depressed by metabolic poisons that inhibit the supply of metabolic energy in the form of ATP. Indeed, in selected instances, ATP serves not as an energy source but as purinergic agonist (see section 7-2), and its role does not involve hydrolysis and phosphorylation. Several ATP-gated channels have been described in cell membranes (see section 5-6); in some instances the compound acts at extracellular sites.

Besides the need for a normal supply of metabolic energy to support active solute transport across the cell membrane, and for the required conditions such as voltage, ATP concentration, ionized Ca levels, or other controlling variables to assure proper gating of solute channels, other factors also contribute to volume control. These include, for instance, the chemical composition of the extracellular medium, the properties of cytoskeleton and extracellular matrix, and the metabolism of specific organic osmolytes. These are synthesized in the cell to prevent its shrinkage in response to external hypertonicity, or they may be broken down or extruded to prevent swelling in hypotonic media (see next section).

8-4 ORGANIC OSMOLYTES

As has been explained, the volume of most cells is primarily determined by the osmolarity of their cytoplasm, and this in turn reflects mostly transmembrane solute fluxes and the metabolism of specific organic osmolytes. The cytoplasmic concentration of these osmolytes in some tissues and species makes up a major fraction of the total osmolarity of cell contents and thereby minimizes cell shrinkage in hyperosmotic surroundings. In marine elasmobranch fish, osmotic water loss to the environment is prevented by raising the osmolarity of extracellular fluid to that of sea water; the difference between the relatively low electrolyte concentrations and the high required solute levels is made up by organic osmolytes such as urea and trimethylamine oxide.

Synthesis or breakdown and loss of organic osmolytes provides a mechanism for cells flexibly to adapt to osmotic insults. Cells in the medulla of the mammalian kidney, for instance, increase their sorbitol content in response to hypertonicity of the interstitium during hydropenia (Burg, 1994).

Quantitatively the chief inorganic osmolyte in most normal cells is K^+. In the absence of metabolism, sodium and chloride leak into cells (see section 8-3), and their contribution to cytoplasmic osmolarity causes cell swelling. The nature of organic osmolytes, in contrast, varies widely in different tissues. Cells from the inner medullary collecting duct of the kidney contain osmotically relevant concentrations of sorbitol, glycerylphosphoryl choline, myoinositol, betaine, and taurine (Burg, 1994). According to Chen and Kempson (1995), free neutral amino acids, including the β-amino acid taurine, polyhydric alcohols like glycerol and sorbitol, and methylamine are the three types of organic osmolytes whose concentrations are found most frequently to vary in osmotically stressed cells.

An additional complication is introduced if these organic compounds not only serve as inert osmolytes but also are pharmacologically active. Taurine, for instance, was observed by Sada, Ban, and Sperelakis (1996) to depress the activity of the Na^+ channel in the membranes of guinea-pig ventricular myocytes.

Swelling or contraction of animal cells under osmotic stress triggers compensatory changes in the concentrations of cellular osmolytes (Chamberlin and Strange, 1989). Such changes are a major factor in the compensatory activity that permits the cell to return toward its normal volume: Regulatory volume decrease in hypotonic media is associated with a decrease in intracellular osmolyte concentration; regulatory volume increase under hypertonic conditions requires an increased concentration of osmolytes in the cells. Need for such regulatory ability is especially critical for cells such as those in the inner renal medulla, the ureters, and the bladder, which for instance in rats may be exposed to extracellular osmolarities ranging from 100 to 3000 mosM.

The metabolism of intracellular osmolytes and their flux across the cell membrane are subject to inhibition by various agents. Pertussis toxin, a known G-protein inhibitor, diminished hypotonic release of sorbitol from inner medullary collecting duct cells (Ruhfus, Tinel, and Kinne, 1996). The contribution of membrane carriers to the turnover of organic osmolytes may be illustrated by the Hg-sensitive taurine channels in polychaete coelemocytes (Preston et al., 1991) and skate hepatocytes (Ballatori and Boyer, 1996) (see section 6), and by the regulatory activity of neutral amino acid transporters in various cell types (Chen and Kempson, 1995).

The role of the neutral amino acid carrier system (system A transporters) in compensatory volume increase of rat renal mesangial cells under hypertonic stress is illustrated in Fig. 8-1 by the stimulation of the uptake of the neutral amino acid methylaminoisobutyrate (Yamauchi et al., 1994) as a function of the hyperosmolarity of the bathing solution.

Over a range of osmolarities ranging from normal to 500 mosM, all compatible with cell survival, compensatory responses are seen to increase with the osmotic challenge. At 600 mosM and above, the challenge proves excessive and no longer elicited a suitable response. Further, the response is depressed by the inhibitors of RNA and protein-synthesis cycloheximide (3.5 μM) and actinomycin D (0.40 μM).

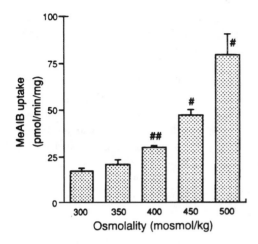

FIG. 8-1. Neutral amino acid uptake in hypertonic media. (From Yamauchi et al., 1994; with permission.)

Although the change in osmolyte uptake ultimately must reflect alterations in membrane function, the effect in this case appears to be induced indirectly by mechanisms involving synthesis of new transporter protein (see section 4-3) and its incorporation into the membrane (membrane traffic; see section 2-5). The importance of osmolyte uptake in permitting cells to compensate for hyperosmolar conditions also is emphasized in the report that survival and growth of MDCK cells in hypertonic media are depressed on inhibition of myoinositol transport with the competitive analogue 2-O,C-methylene-myoinositol (Kitamura et al., 1997).

Together with the stimulation of amino acid uptake, the cellular contents of almost all of the amino acids measured by Yamauchi et al. (1994) increased two-fold or more in hypertonic media. The difference in the free amino acid concentration between control and hypertonic conditions was 106 nmol/mg protein, corresponding to an osmolarity difference in total cell water of around 40 mosM. Whether the 106 nmol can fill the whole osmolarity gap between cells in isotonic and hypertonic media depends on the intracellular volumes of distribution of the amino acids. The implicit assumption that the intracellular volume of distribution of amino acids equals the whole cell water has not been tested.

8-5 MEMBRANE MECHANISM OF VOLUME HOMEOSTASIS

Preceding sections have emphasized the importance of membrane-associated functions in permitting cells to control their composition and volume; loss of this control is associated with serious cell malfunction. The close relationship between cell function and cell volume often has been emphasized, for instance in the work of Haussinger, Lang, and Gerok (1994).

The question next arises of the mechanism of volume homeostasis: How are the various membrane functions tied into the negative feedback loops that determine the final steady-state volume of a cell? Several membrane functions, indeed, are sensitive to volume expansion or contraction. An expansion of cell volume, for instance, stimulates stretch-sensitive solute channels in the membrane. The stretch response is elicited on cell swelling in hypotonic media (Hoffmann and Simonsen, 1989); compensatory membrane changes, which in this case lead to increased solute efflux from cells, subsequently tend to return the cell volume towards normal. Cell swelling also may influence the rate of a membrane process by altering cell geometry and thereby the surface area of exposed membrane actively participating in, for example, solute exchange.

Membrane constituents and processes participate both as sensors and as compensators in negative feedback loops regulating cell volume. To illustrate such a loop consider a cell acutely swollen in a hypotonic medium; swelling is associated with stretching of the membrane. Present in this membrane are a variety of stretch-sensitive solute channels, whose activity has been measured both in intact cells and in membrane patches (Chamberlin and Strange, 1989; Christensen, 1987) (see section 3-7). Swelling thus can trigger changes in membrane permeability, and these tend to counteract the effects of hypotonicity and thus reverse the increase in the cell volume.

An example of such compensation is provided by the volume-activated K^+ and Cl^- channels in human lymphocytes (Sarkadi, Mack, and Rothstein, 1984). Activation of these channels in hypotonically swollen cells permits loss of KCl and thereby leads to regulatory volume decrease. Similarly, mouse lymphoma cells swollen in hypotonic media lose potassium because of a transient increase in K permeability (Roti Roti and Rothstein, 1973); no volume compensation occurs in absence of K gradients in K-free media, or in presence of ouabain. Efflux of chloride through the stretch-activated chloride channel in swollen cells is accompanied by increased efflux of the organic osmolytes taurine and myoinositol (Hand, Morrison, and Strange, 1997).

Other volume-sensitive membrane channels and transporters from various cells include calcium and chloride channels, a chloride–bicarbonate exchanger, and cotransporters for Na-Cl and Na-K-2Cl. Osmotic regulation of mRNA for the Na-myoinositol cotransporter has been described in endothelial and neural cells (Wiese et al., 1996). Additional sensors capable of stimulating cellular responses to extracellular osmolarity changes are the osmoreceptors described in yeast cells (see section 4-5). They are believed to trigger an osmosensing and mitogen-activated protein kinase cascade that controls the rate of glycerol synthesis. Because an outer cell wall prevents osmotic swelling of yeast cells, however, the sensors in this case presumably are responding to increases in turgor pressure resulting from exposure to hypotonic media, rather than to cellular volume changes and membrane stretching.

The feedback loop acting at the membrane to maintain normal volume of eucaryotic cells is likely, as mentioned, to be mediated by stretch receptors. The influence of cell volume on membrane channels, however, need not represent a simple stretch response. For instance, Oike, Droogmans, and Nilius (1994) found chloride channels in human endothelial cells that are not activated by cell swelling in the absence of a critical level of ATP in the cell. Similarly, the volume-activated and mercury-sensitive taurine channel of skate hepatocytes requires ATP to remain in the open state (Ballatori, Simmons, and Boyer, 1994). Jackson, Morrison, and Strange (1994) described a relatively nonselective organic anion–osmolyte channel in rat glioma cells that is activated by cell swelling and is regulated by the nonhydrolytic binding of ATP. Lowering of ATP levels, for instance by adding azide or DNP, inhibits efflux through this channel. This represents a typical indirect effect on membrane function, elicited by toxicants interfering with ATP synthesis (see Chapter 5). The nucleotide here appears to act not as energy-rich substrate but rather as a purinergic agonist.

The linkage of membrane function to cell volume may involve other as-yet-unidentified mechanisms. Thus, a possible role has been proposed for the cytoskeleton as a sensor in volume control (Hoffmann and Simonsen, 1989). Motais, Guizouarn, and Garcia-Romeu (1991) suggested that ionic strength may influence efflux of osmotically important amino acids from trout erythrocytes as a result of its influence on the cytoskeleton. Cell shape and function also may be influenced by the extracellular matrix. What all these findings point to is the conclusion that volume maintenance is a highly integrated process correlated with function of the cell membrane. This explains why any attack on the functional integrity of cells is likely to be reflected in loss of normal volume control and compensatory volume regulation. It also accounts for the finding that toxicants can affect cellular volume regulation by direct reaction with membranes; this is considered further in the next section.

8-6 EFFECTS OF TOXICANTS ON VOLUME REGULATION

Because any interference with basic cell function is likely to result in volume changes, it is not surprising to find that cytotoxicity generally is associated with cell swelling. Indeed, swelling of cells (and mitochondria) is routinely used as a monitor for detecting cytotoxic effects, whether in vivo or in isolated tissues or cells. The association between toxicity and changes in cell volume (swelling in isotonic solution) can be illustrated by many instances; two are cited here. The first example comes from the work of Dudley, Svoboda, and Klaassen (1984), who described the swelling of hepatocytes from Cd-poisoned rats. The second example consists of the hemolytic action of arsine (AsH_3). This toxicant exerts an early effect on the ability of erythrocytes to maintain normal Na and K levels (Winski et al., 1997); as a result, cells swell and hemolyze. Profound changes are observed in the membrane structure of exposed cells. The fact that arsine does not alter significantly levels of ATP supports the hypothesis of a direct action of the poison on membrane function.

Generally, changes in cell (or mitochondrial) volume as seen under the microscope provide qualitative evidence for cell malfunction. More quantitatively, Kleinzeller and Cort (1957) attempted to evaluate the cytotoxic effects of Hg on renal cortex by the measurement of water uptake in slices. A difficulty in this approach, however, arises from the fact that Hg increases the water content of the tubular lumina by inhibiting normal sodium and water absorption (see section 3-3); as a result the increased water content of the slices in presence of the metal can be attributed only in part to cellular edema; the magnitude of residual tubular fluid volumes in slices is difficult to estimate. Numerous other examples, however, can be cited of the relationship between cell volume and cytotoxicity in various tissues.

Since the earlier contributions from Kleinzeller's laboratory, considerable further attention has focused on the mechanism and sites of action of Hg (and its monomethyl compound) on cell volumes. As expected from their high affinity for proteins in general, Hg^{++} or its organic derivative, monomethylmercury, depress the activity of channels for water (aquaporins) and a variety of solutes involved in volume regulation at the cell membrane; prime among these are Na and taurine. Vitarella, Kimmelberg, and Aschner (1996), for instance, reported that methylmercury inhibits volume regulation in astrocytes. In the absence of sodium, however, the mercury compound did not affect the compensatory shrinkage of cells swollen in hypotonic media (regulatory volume decrease). The effect of methylmercury on volume homeostasis in astrocytes can be explained readily on the basis that cell shrinkage is normally associated with the active and Hg-sensitive extrusion of salt and water (see section 8-3).

As already pointed out, it may be difficult in many cases to decide whether such volume effects of mercury (or other toxic agents) directly involve an interaction between membrane solute channels and the toxicant. Ballatori and Boyer (1996), in their work on the mercury inhibition of cell volume regulation in skate hepatocytes, described how 50 μM $HgCl_2$ inhibits the regulatory volume decrease after cells have been swollen by sudden exposure to hypotonic solutions (Fig. 8-2). At a slightly lower concentration (4 μM), Hg abolished the activity of the membrane channel for taurine, a critical osmolyte in volume regulation in these cells, without affecting the intracellular ATP levels or ATP/ADP ratios. This finding is compatible with a direct metal effect on membrane function relating to extrusion of osmolytes (Na, K, and taurine) from the shrinking cell, rather than with an indirect effect mediated by inhibition of energy metabolism.

Replacement of Na with K in Fig. 8-2c abolishes compensatory activity. Extrusion of Na under these conditions obviously can play little part in volume compensation. One explanation for the absence of compensatory activity is that neither K nor taurine extrusion occurred in K-Ringer's. The reasons for the slower volume shrinkage in choline Ringer's solution (see Fig. 8-2b), and the absence of Hg effect are not clear.

Even in isotonic Na Ringer's, 50 μM $HgCl_2$ causes the cells to swell, their volume doubling in 120 minutes; at the same time their sodium content almost triples. The gain in Na greatly exceeds the loss of K, reflecting the net gain of osmotically active solutes responsible for the inability of these cells to compensate for swelling under these conditions.

The sensitivity of taurine channels to mercurials also has been described in polychaete coelomocytes (Preston et al., 1991). The β-amino sulfonic acid is the major organic osmolyte in these cells, and volume regulation is intimately related to taurine fluxes. Table 8-1 describes these results.

Mercury inhibition of taurine transport is reversed readily by dithiothreitol and L-cysteine; somewhat larger monothiols (acetyl cysteine, glutathione) exert little inhibition. This finding is interpreted in terms of Hg acting not at an easily accessible metal-binding site on the outer surface of the cell membrane, but rather at a site more deeply buried in the membrane, or on the inner aspect of the membrane.

TABLE 8-1. *Mercury inhibition of taurine transport and its reversal in red blood cells from glycera*

	Inhibition, %
PCMBS (1 mM)	64
DTT (20 mM)	10
PCMBS, followed by DTT	28

The cells were preincubated for 1 minute with saline (control) or p-chloromercuribenzenesulfonate (PCMBS); cells were then exposed to saline or dithiothreitol (DTT) for 10 minutes, before being allowed to take up taurine (initial concentration 1 mM) for 5 minutes.
Data from Preston et al. (1991).

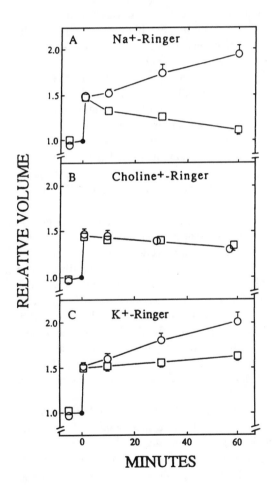

FIG. 8-2. Effect of Hg on regulatory volume decrease. Skate hepatocytes were swollen by a 40% dilution of the suspending Ringer's solution with H_2O, in absence (*squares*) or presence (*circles*) of 50 μM $HgCl_2$. In (*a*) note the inhibition of normal conpensatory shrinkage by Hg. (*b*) NaCl is replaced by choline chloride. (*c*) NaCl is replaced by KCl. (From Ballatori and Boyer, 1996; with permission.)

Large volume changes have been observed in freshly isolated lymphocytes from rat blood within minutes of exposure to Hg. Thus, 0.1 mM Hg in 10 minutes led to a greater than 50% increase in the volume of the cytoplasm, without, however, a simultaneous swelling of subcellular organelles (Foulkes and Blanck, 1994). Unfortunately, the concentration of active Hg at the intracellular membranes under these conditions is not known, so that no conclusion can be drawn on the relative Hg sensitivities of cell and mitochondrial membranes. The important conclusion must be reemphasized, however, that swelling of cytoplasm and of mitochondria are separate processes.

In various tissues, inorganic mercury also potently reduces bulk flow of water (see section 5-3). It inhibits, for instance, the water channels involved in the shrinkage of osmotically swollen cells, as shown by Preston et al. (1992). The activity of the erythrocyte water channel CHIP28 expressed in *Xenopus* oocytes was about 60% inhibited by 0.3 mM $HgCl_2$. Although this is admittedly a high concentration of the metal, and the effect presumably is completely nonspecific, it is essentially completely reversed by addition of 5 mM mercaptoethanol. Many other chemicals besides Hg also can alter cell permeabilities and thereby cell volumes; this emphasizes again the close correlation between cytotoxicity and cell swelling.

There is suggestive evidence also for the inverse effect of a toxicant, whereby cell swelling may become the cause rather than the effect of membrane malfunction. The existence of stretch-sensitive channels is cited previously. Foulkes and Blanck (1991) reported that treatment of rabbits with Cd leads to equal fractional depression of active basolateral cycloleucine accumulation in renal cortex and of its passive basolateral extrusion. As pointed out in section 5-6, such observations can be explained by a direct interaction of the metal with an amino acid transporter mediating fluxes of its substrate in either direction. Alternatively, one might invoke a primary action of the metal on cell swelling, with consequent changes in membrane geometry and stretching; stretch receptors might influence secondarily the rate of amino acid transport.

Effects of toxicants and drugs specifically on membrane processes involved in compensatory cell swelling or shrinkage also have been demonstrated. Reference already has been made to the Hg inhibition of bulk water flow out of osmotically swollen cells (Preston et al., 1992). Kennedy (1994) observed that inhibition of K flux by the K-channel blocker Ba depresses regulatory volume decrease in human retinal cells exposed to hypotonic solutions. The inhibition of regulatory volume decrease in human lymphocytes by the stilbene derivative DIDS, a specific anion channel blocker (Sarkadi, Mack, and Rothstein, 1984), points to the involvement of anion channels in volume control. The role of organic osmolyte–anion channels in volume control of rat glioma cells (Jackson, Morrison, and Strange, 1994) has been referred to previously.

In addition to toxic effects on solute permeability of the membrane, and in particular on the flux of one of the many specific organic osmolytes that have been identified in various cell types (see section 8-4), toxicants theoretically also may affect compensatory volume regulation by, for example, depressing osmolyte synthesis. This would diminish the ability of cells gradually to return toward normal volume in hypertonic solutions. Evidence for specific inhibitors of this kind is limited, however.

The effect of toxicants on tissue swelling so far has been discussed in terms of volume changes of cell or cytoplasm. In passing, it should be recalled that swelling of tissues in vivo also may result, not from expansion of the cellular compartment but from interstitial edema following, for example, a loss of plasma oncotic pressure. Whether cellular or interstitial, swelling might interfere with blood circulation and thereby cellular O_2 supply and metabolism.

8-7 SUMMARY

Most mammalian cells can maintain to a large extent their normal volume by continuously pumping out sodium; this ion, together with chloride, leaks down the electrochemical activity gradient into the cell. Acute swelling of cells in vivo, or under isotonic conditions in vitro, therefore commonly can be ascribed to changes in membrane function, and more specifically to inhibition of active sodium transport. The inhibition may be exerted directly at the membrane or may be an indirect consequence of depression of energy metabolism or other basic cell functions. In full agreement with this interpretation, the action of toxic agents often is reflected in changes in volume homeostasis. Swelling of cells under isotonic conditions thus can provide a sensitive biomonitor of cytotoxicity.

Similarly, if cells have swollen passively or contracted in nonisotonic media, their ability to shrink or swell in order to compensate for osmotic volume changes, and thus to return their volume toward normal, also is sensitive to the action of many toxic agents. In part, compensation here involves changes in the metabolism of cellular osmolytes. As in the case of normal volume homeostasis, however, such compensatory or adaptive changes also depend on the control of solute permeability at the cell membrane. The fact that control of solute fluxes across the membrane thus represents a critical contributor to cell volume regulation makes the inhibition of volume homeostasis a significant area in membrane toxicology.

The interpretation of data on cell swelling in some published work, and their toxicologic significance, is at times obscured by the failure to distinguish between volume changes of cytoplasm or of subcellular organelles. This can lead to inappropriately simplified conclusions, as volume changes of, for example, mitochondria may differ from those of cytoplasm. Indeed, available evidence suggests that volume control of cytoplasm and of subcellular organelles are independent processes and may be differently affected by toxicants.

A further unresolved difficulty in studies of cellular volume changes is that of how to estimate the extracellular spaces in some tissues. This volume must be known so that cellular swelling can be fully distinguished from, for example, interstitial edema. A similar problem arises specifically in renal tissue containing undetermined volumes of tubular fluid. This fluid may contain unknown concentrations of extracellular markers like inulin; as a result, subtraction of the inulin space as normally calculated from total tissue water cannot yield adequate values for intracellular volume. It would be useful if swelling or contraction of cells could be defined always in terms of the volumes actually involved.

Even without such refinements, however, the conclusion is evident that toxicants can severely affect the mechanism of cell-volume homeostasis, a process strongly dependent on membrane function, and that changes in volume homeostasis provide a sensitive biomonitor of cytotoxicity.

REFERENCES

Ballatori, N., and J.L. Boyer. 1996. Disruption of cell volume regulation by mercuric chloride is mediated by an increase in sodium permeability and inhibition of an osmolyte channel in skate hepatocytes. *Toxicol. Appl. Pharmacol.* 140:404–410.

Ballatori, N., T.W. Simmons, and J.L. Boyer. 1994. A volume-activated taurine channel in skate hepatocytes: Membrane polarity and role of intracellular ATP. *Am. J. Physiol.* 267:G285–G291.

Burg, M.B. 1994. Molecular basis for osmoregulation of organic osmolytes in renal medullary cells. *J. Exp. Zool.* 268:171–175.

Chamberlin, M.E., and K. Strange. 1989. Anisosmotic cell volume regulation: A comparative view. *Am. J. Physiol.* 257:C159–C173.

Chen, J.G., and S.A. Kempson. 1995. Osmoregulation of neutral amino acid transport. *Proc. Soc. Exp. Biol. Med.* 210:1–6.

Christensen, O. 1987. Mediation of cell volume regulation by Ca^{++} influx through stretch-activated channels. *Nature* 330:66–68.

Dudley, R.E., D.J. Svoboda, and C.D. Klaassen. 1984. Time course of cadmium-induced ultrastructural changes in rat liver. *Toxicol. Appl. Pharmacol.* 76:150–160.

Foulkes, E.C., and S. Blanck. 1991. Cadmium inhibition of basolateral solute fluxes in rabbit renal tubules and the nature of cycloleucine uptake. *Toxicol. Appl. Pharmacol.* 108:150–156.

Foulkes, E.C., and S. Blanck. 1993. Volume of renal cortical cytoplasm in rabbits and basolateral transport gradients of cycloleucine and PAH. *Proc. Soc. Exp. Biol. Med.* 202:302–306.

Foulkes, E.C., and S. Blanck. 1994. 3-O-Methylglucose as probe of cytoplasmic volume. *Life Sci.* 54:439–444.

Hand, M., R. Morrison, and K. Strange. 1997. Characterization of volume-sensitive osmolyte efflux and anion current in *Xenopus* oocytes. *J. Membr. Biol.* 157:9–16.

Haussinger, D., F. Lang, and W. Gerok. 1994. Regulation of cell function by the cellular hydration state. *Am. J. Physiol.* 267:E343–E355.

Hoffmann, E.K., and L.O. Simonsen. 1989. Membrane mechanisms in volume and pH regulation in vertebrate cells. *Physiol. Rev.* 69:315–382.

Jackson, P.S., R. Morrison, and K. Strange. 1994. The volume-sensitive organic osmolyte-anion channel is regulated by non-hydrolytic ATP binding. *Am. J. Physiol.* 267:C1203–C1209.

Kennedy, B.G. 1994. Volume regulation in cultured cells derived from human retinal pigment epithelium. *Am. J. Physiol.* 266:C676–C683.

Kitamura, H., A. Yamauchi, T. Nakanishi, Y. Takamitsu, T. Sugiura, A. Akagi, T. Moriyama, M. Horio, and E. Imai. 1997. Effects of inhibition of myo-inositol transport on MDCK cells under hypertonic environment. *Am. J. Physiol.* 272:F267–F272.

Kleinzeller, A., and J.H. Cort. 1957. The mechanism of action of mercurial preparations on transport processes and the role of thiol groups in the cell membrane of renal tubular cells. *Biochem. J.* 67:15–24.

Lang, F., and D. Haussinger. 1993. *Interactions of cell volume and cell function.* Berlin: Springer Verlag.

Leaf, A. 1959. Maintenance of concentration gradients and regulation of cell volume. *Ann. N.Y. Acad. Sci.* 72:396–404.

Miller, D.S., D.E. Stewart, and J.B. Pritchard. 1993. Intracellular compartmentation of organic anions within renal cells. *Am. J. Physiol.* 264:R882–R890.

Motais, R., H. Guizouarn, and F. Garcia-Romeu. 1991. Red cell volume regulation: The pivotal role of ionic strength in controlling swelling-dependent transport systems. *Biochim. Biophys. Acta* 1075:169–180.

Oike, M., G. Droogmans, and B. Nilius. 1994. The volume-activated chloride current in human endothelial cells depends on intracellular ATP. *Pflugers Arch.* 427:184–186.

Preston, G.M., T.P. Carroll, W.B. Guggino, and P. Agre. 1992. Appearance of water channels in *Xenopus* oocytes expressing red cell CHIP28 protein. *Science* 256:385–387.

Preston, R.L., S.J. Janssen, S. Lu, K.L. McQuade, and L. Beal. 1991. Organic and inorganic mercurial inhibition of taurine transport by the coelemocytes of the marine polychaete, *Glycera dibranchiata. Bull. Mt. Desert Island Biol. Lab.* 30:72–74.

Roti Roti, L.W., and A. Rothstein. 1973. Adaptation of mouse leukemia cells (L5178Y) to anisotonic media: I. Cell volume regulation. *Exp. Cell Res.* 79:295–310.

Ruhfus, B., H. Tinel, and R.K.H. Kinne. 1996. Role of G-proteins in the regulation of organic osmolyte efflux from isolated rat renal inner medullary collecting duct cells. *Pflugers Arch.* 433:35–41.

Sada, H., T. Ban, and N. Sperelakis. 1996. Kinetic mechanisms of Na^+ channel depression by taurine in guinea pig ventricular myocytes. *Japn. J. Pharmacol.* 71:147–159.

Sarkadi, B., E. Mack, and A. Rothstein. 1984. Ion events during the volume response of human peripheral blood lymphocytes to hypotonic media. *J. Gen. Physiol.* 83:497–512.

Vitarella, D., H.K. Kimmelberg, and M. Aschner. 1996. Regulatory volume decrease in primary astrocyte cultures: Relevance to methylmercury neurotoxicity. *Neurotoxicology* 17:117–123.

Wiese, T.J., J.A. Dunlap, C.E. Conner, J.A. Grzybowski, W.L. Lowe, and M.A. Yorek. 1996. Osmotic regulation of Na-myoinositol cotransporter mRNA level and activity in endothelial and neural cells. *Am. J. Physiol.* 270:C990–C997.

Winski, S.L., D.S. Barber, L.T. Rael, and D.E. Carter. 1997. Sequence of toxic events in arsine-induced hemolysis in vitro: Implications for the mechanism of toxicity in human erythrocytes, *Fundam. Appl. Toxicol.* 38:123–128.

Yamauchi, A., A. Miyai, K. Yokoyama, T. Itoh, T. Kamada, N. Ueda, and Y. Fujiwara. 1994. Response to osmotic stimuli in mesangial cells: Role of system A transporter. *Am. J. Physiol.* 267:C1493–C1500.

Subject Index

A

Absorption, toxicant
  pulmonary, 103
  steps of, 91
Acetaldehyde, 75
Acylcarnitine, 77–78
Adaptation, cell, 127–128
Adaptive up-regulation, 128
Adrenal insufficiency, 61
Adriamycin, 68, 79, 124
Agatoxin, 71
L-Alanine, 67
Alcohols, 72
Amino acids, renal absorption of, 13
Amino-acid carrier system, 136
4-Aminopyridine, 71
Ammonia, 90
Anesthetics, 64
  local, 6–7, 72
Annexins, 47
Anthracene-9-carboxylate, 71
Anthracyclines, 68
Antiarrhythmic drugs, 64
Antibiotics, beta-lactam, 108–109
Antidiuretic hormone, 54
Antimycin A, 78
Aquaporin, 54
Aquaporin channels, 54
Aqueous pores, 53–54, 132
Arsenite, extrusion of, 120–121
Artifacts
  experimental, 21–22
  in small intestine everted sacs,
    32–33
ATP, 72
  P glycoprotein and, 126–127
Autoimmune lesions, 83
Azide, 61, 69

B

Bacterial toxins, 64, 72 (*See also specific bacteria*)
Bafilomycin A, 71
Basolateral transport, 25

Bee venom, 71
1,2,3-Benzenetricarboxylate, 78
Beta-lactam antibiotics, renal tubule uptake
  of, 108–109
Blood–brain barrier, vs. placenta, 111
Blood–brain barrier toxicant transport,
  109–111
Brush border
  in slices, 27–28
  in solute transport, 30–31
Bungaratoxin, 71
N-Butanol, 64
Butoxyacetic acid, 66

C

Cadmium (*See also heavy metals*)
  on ATP-requiring Ca transport, 99–100
  on basolateral solute transport, 75–76
  binding sites of, 66
  calcium in uptake of, 104
  cell swelling from, 76, 139
  EDTA on uptake of, 96, 98
  electric charge changes with, 65–66
  on endothelin receptors, 74
  on glucose transport, 73
  hepatic uptake of, 103–105
  immunotoxic action of, 82–83
  on jejunal brush border, 91–92
  on L-alanine uptake, 73
  membrane uptake of, 96–99
  membrane-bound, compartmentalization
    of, 98–99
  multiple effects of, 63
  nephrotoxicity of, 61–62
  pulmonary absorption of, 103
  renal transport of, 105–108
  time-dependent toxicity of, 76
Cadmium-metallothionein, 10–11
Cadmium-resistance protein, 121
Calcium, in osmolarity, 7
Calcium channel, 53
  in cadmium uptake, 104
Calcium channel blockers, 79
Carbon tetrachloride, 79

K

Kidney (*see* Renal)
Kinetic models, of intestinal absorption,
32

L

Lactate dehydrogenase, 8
Laminin, 83
Lead, 64, 73 (*See also heavy metals*)
  on antigen handling, 82
  brain uptake of, 111
  immunotoxic action of, 82
  on membrane proteins, 47–48
Lesions, membrane, 60–61
  direct vs. indirect, 61–63
  indirect, 74–77
Lindane, 75
Lipid peroxidation, toxicants on, 8
Lipophilicity, in drug action and toxicology,
  12
Lipoprotein matrix, 47–51 (*See also
  proteins, membrane*)
Lipoprotein structure, toxicants on, 8
Liposomes, as drug carriers, 12
Liver (*see* Hepatic)
Lysosomal membrane, 77
  damage to, 11

M

Maitotoxin, 71
Malondialdehyde, 67
Malonyl-coenzyme A, 67, 77–78
Mastoparan, 71
Mechanical stimuli, 52
Mechanical strength, 131–132
Medalin, 46
Melittin, 71
Membrane
  complexity of, 1
  damage to, 1
  history of study of, 1
Membrane patches, 35–37
Membrane-associated proteins, 47–51 (*See
  also proteins, membrane*)
Membranes, biological, 5–16 (*See also
  specific aspects*)
  functions of, 5
  functions of, adaptive changes in,
    117–128
  homeostasis and, 6–7

protective and sequestering roles of, 8–14
  (*See also permeability barrier*)
  synthesis of and traffic in, 15–16
  as target sites, 7–8
  toxicant distribution and, 14–15
Mercurials (*see* Dimethyl mercury;
  Mercury; Methylmercury; Monomethyl
  mercury)
p-Mercuribenzenesulfonate, 73
Mercuric mercury, 7–8
  blood–brain barrier exclusion of, 12
  on membrane enzymes, 51
Mercury, 62–63, 66, 91 (*See also dimethyl
  mercury; heavy metals; methylmercury;
  monomethyl mercury*)
  on amino acid absorption, 69
  on ATPase, 73
  blood–brain barrier on, 110–111
  on cell volume, 139–142
  electric charge changes with, 65–66
  erythrocyte uptake of, 99
  on glucose transport, 68–70
  immediate effects of, 76–77
  immunotoxic action of, 82
  multiple membrane effects of, 62–63
  nephrotoxicity of, 83
  on solute channels, 53
  at synaptic membranes, 69
  on toxicant extrusion, 122–123
Mersalyl, 78
Messengers
  first, 51
  second, 52
Metabolic inhibitors, 74–75
Metal efflux proteins, 121
Metallothionein, 128
  Cd trapping of, 10–11
Metals (*See also heavy metals; specific
  metals*)
  electric charge changes with, 64–65
  vesicle transport of, 36–37
3-O-Methylglucose (3OM), 80–81
Methylmercury, 63
Mitochondria
  toxicants on, 2
  volume homeostasis in, 133
Mitochondrial membranes, 77–78
Monensin, on pinocytosis, 16
Monomethyl mercury, 93
  blood–brain barrier transport of,
    110–111
  intestinal absorption of, 101–102